MW00824295

OCEAN LIFE IN THE TIME OF DINOSAURS

Nathalie Bardet
Alexandra Houssaye
Stéphane Jouve
Peggy Vincent

Illustrated by
Alain Bénéteau

Translated by
Mark Epstein

Princeton University Press
Princeton and Oxford

CONTENTS

PROLOGUE

> The 17th century itself would have loved these mountains, which could be seen as gigantic and harmonious constructions. And this splendid "Painted desert" where plesiosaurs and pterodactyls have left traces of their passage.
>
> —Maurice Ravel, letter to Roland Manuel, April 14, 1928, following his visit to the Grand Canyon
>
> (Manuel Cornejo, *Maurice Ravel, l'integrale. Correspondances (1895–1937), ecrits et entretiens,* Paris, Le Passeur, 2018, Letter 2171, p. 1071 [translated])

Fossil. The mere word kicks most people's imagination into high gear. It mostly evokes dinosaurs, of course—the stars of the Mesozoic era. And yet, because those famous reptiles are already the focus of many other books, in this book we will discuss different but no less fascinating kinds of animals: the ones that many people refer to incorrectly as "marine dinosaurs."

To explain how these reptiles—which were contemporaries of dinosaurs and could be just as imposing—were completely different from their terrestrial relatives, we will take you on a voyage that begins roughly 300 million years ago (abbreviated as 300 Ma), on an Earth considerably different from the one you know today. It was around that time that reptiles started to make significant returns to the aquatic life of their (and your) ancestors.

During the Mesozoic (252–66 Ma), considered the golden age of reptiles, reptiles predominated in all ecosystems and occupied extremely varied ecological niches—on land, following the example of the dinosaurs; in the air, with the pterosaurs; and in the oceans. Yet, about a hundred million years after four-legged vertebrates (tetrapods) first left the water and colonized the land, which they did in the Late Devonian (380–360 Ma), a certain number of lineages followed the opposite path. That is to say that, long before the appearance of the first dinosaurs, several groups of reptiles, all of which originated from terrestrial ancestors, secondarily and independently adapted to the aquatic environment. The kinds of pressures from natural selection that led to this (re)colonization of the aquatic environment are still being researched.

Whatever the reasons for this secondary adaptation to the aquatic environment, it is considered one of the most important events in the evolution of the vertebrates. Although the reptiles that took this path are broadly categorized as "marine reptiles," they varied significantly in their anatomy, their size, and their habits. Each type had undergone considerable anatomical and physiological changes from its land-dwelling forebears, changes that allowed it to survive in a marine environment and maintain its competitiveness. They included modifications that provided better hydrodynamics, balance, locomotion, thermoregulation, reproduction, nutrition, and respiration.

Among the marine reptiles that the broader public is familiar with (and to which we will return repeatedly in this book), ichthyosaurs, plesiosaurs, and mosasaurs were at the top of the food chain. They are emblematic of Mesozoic marine reptiles. Among their entourage were a stream of other reptiles, sometimes less well known but equally surprising. Some belonged to groups that have no modern representatives, while others are—albeit with significant differences—still found today. Among the latter are crocodiles (order Crocodilia); lizards, snakes, and other scaly reptiles (order Squamata); and tortoises (order Testudinata). But unlike their living representatives, the marine reptiles of the Mesozoic ranged in size from small to gigantic and demonstrated a remarkable array of adaptations, both morphological and ecological. We shall therefore proceed to tell their history.

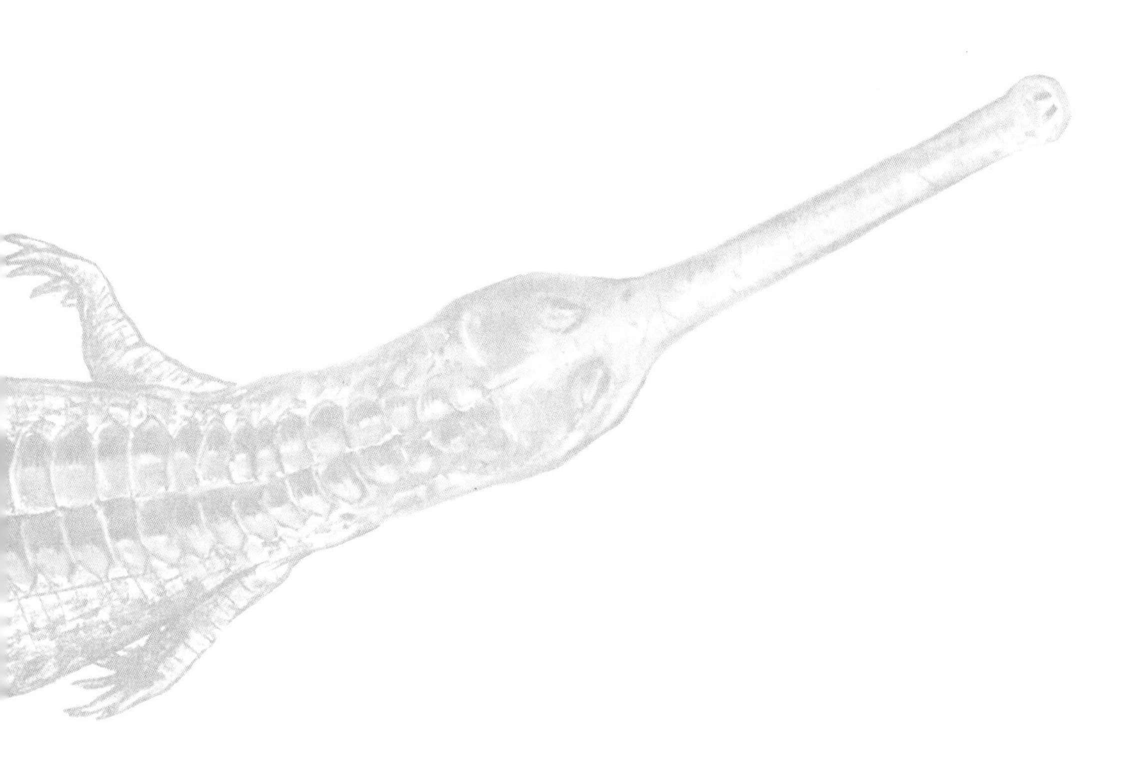

CHAPTER 1

SETTING THE SCENE

Tectonics is the opium of the landscape.

— Sylvain Tesson, (*Sur les chemins noirs*, Paris, Gallimard, 2016 [translated])

To best depict a landscape, I first have to discover its geological assets.

—Paul Cézanne, letter to Joaquim Gasquet
(Joaquim Gasquet, *Cézanne*, Paris, éditions Bernheim-Jeune, 1923, p. 83 [translated]).

Today, reptiles (including birds, which, from an evolutionary point of view, belong to this group) represent the most diverse group of **amniotic** vertebrates, with more than twenty-one thousand species described so far. And yet, despite this diversity and the immensity of Earth's seas and oceans (with their thousands of islands and thousands of miles of coastline), relatively few kinds of reptiles are at home in salt water (fig. 1.1 and fig. 1.2). As many as 250 species at least occasionally swim in the sea, but this is still a tiny number compared with the more than 23,000 species of teleosts (the order Teleostei includes most modern bony fish) that live there.

What do today's marine reptiles look like?

Modern Marine Reptiles

Sea Snakes

Snakes, along with turtles, are the reptiles that exhibit the most advanced adaptations to the marine environment today. Different varieties of sea snakes—more than eighty species and subspecies, all in family Elapidae—live in fresh water, brackish (somewhat salty) water, and salt water. They are found mainly in the tropical and subtropical areas of the Indian and Pacific Oceans.

Sea snakes' most visible morphological characteristic is their laterally flattened tail, which they use for propulsion. Sea snakes are **viviparous**, meaning they give birth to young that are already fully formed (they do not lay eggs)—except for the genus *Laticauda*, the only sea snake that returns to land to lay its

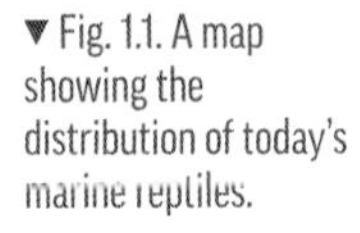

▼ Fig. 1.1. A map showing the distribution of today's marine reptiles.

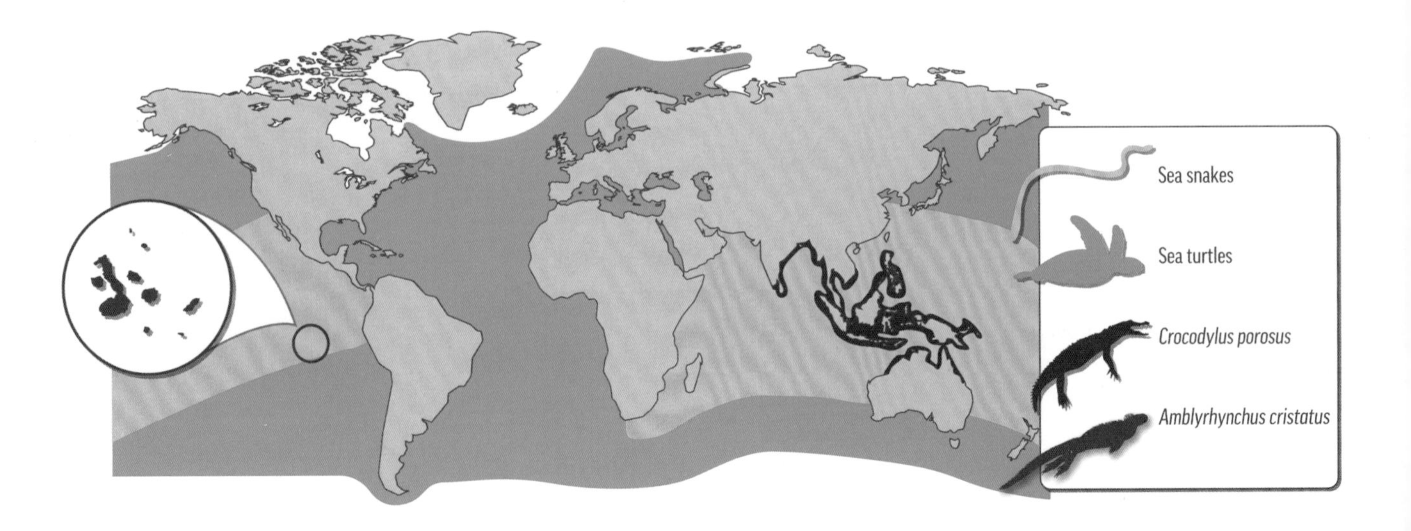

▲ Fig. 1.2. Some examples of today's marine reptiles: (a) the snake *Hydrophis platurus* (family Elapidae); (b) the leatherback turtle *Dermochelys coriacea* (family Dermochelyidae); (c) the Galapagos marine iguana *Amblyrhynchus cristatus* (family Iguanidae); and (d) the saltwater crocodile *Crocodylus porosus* (family Crocodylidae).

eggs. All these serpents are known for the toxicity of their venom. Some, like *Hydrophis schistosus* (about 1.3 meters long), are very aggressive, with venom much more deadly than a cobra's. Still, many sea snakes are relatively mild-mannered, and they do not always inject a considerable amount of venom when they bite. Most live near coastlines, and only *H. platurus* is **pelagic** (can swim in the open ocean) and is geographically widespread.

Sea Turtles

Sea turtles make up the second most diverse group of reptiles that inhabit the sea. And yet they number only seven species, belonging to two families (Cheloniidae and Dermochelyidae). The best known, without a doubt, is the leatherback sea turtle (*Dermochelys coriacea*). Up to 2 meters long and weighing between 250 and 700 kilograms, it is the largest turtle in existence.

Sea turtles live mostly along tropical coastlines, but they can migrate over long distances, using Earth's magnetic field to orient themselves in the open sea. Their remarkable ability to navigate, which is still only very poorly understood, helps them return again and again from the vastness of the oceans to the very same beach on which they were born, in order to lay their eggs. Today, sadly, most sea turtles are endangered by human activities, to the extent that certain species, such as the leatherback turtle and the hawksbill sea turtle (*Eretmochelys imbricata*), are at risk of soon becoming extinct.

A Lizard and a Crocodile

Although no less than half the species of reptiles alive today are lizards, lizards have no representative in the marine environment. Only the "marine" Galapagos iguana (*Amblyrhynchus cristatus*) dares venture into the ocean. This animal, which can grow to

1.2 meters long, propels itself in the water using its laterally compressed tail. Although the marine iguana spends most of its time on land, where it basks in the sun and lays its eggs, it searches for food (mostly marine algae) in the sea.

In some ways crocodiles are in the same situation as the lizards; only one of the roughly twenty species of crocodiles, the estuarine crocodile (*Crocodylus porosus*), is considered unambiguously marine. Also called the saltwater crocodile, it is frequently observed at sea but lives principally in coastal habitats, rivers, deltas, and marshy areas of southeast Asia and Oceania. Sometimes exceeding 7 meters in length, it is the largest crocodile and one of the largest living reptiles. Males compete fiercely to occupy the rivers where the females come to mate. Young adult males are often chased away by the stronger "veterans" and, therefore, are regularly spotted in the open ocean, sometimes dozens of miles from the coast, searching for new territory to colonize. Although estuarine crocodiles spend a good portion of their time at the beach, they can stay at sea for several days, using the currents to navigate.

Although they are fascinating in many respects, today's marine reptiles, consisting of only a few small groups and with relatively restricted geographic ranges, represent a very modest sampling when compared with the extraordinary variety of reptiles around the world in the waters of the Mesozoic. Let us, then, move to the Mesozoic!

The Waltz of the Continents during the Mesozoic

The Mesozoic era (from the Greek *meso*, meaning "middle," and *zoon*, meaning "living being") started 252 million years ago and came to an end about 66 million years ago. During those 186 million years, both Earth itself and life on the planet experienced some profound changes. The Mesozoic is divided into three periods, the Triassic, the Jurassic, and the Cretaceous, which are themselves divided into epochs (e.g., Early Cretaceous and Late Cretaceous) and then ages.

If we go back to the beginning of the Triassic, almost all the land masses on Earth are conjoined, forming a supercontinent called Pangaea (fig. 1.3). In the east, an ocean, the Tethys, cuts into this continent, while

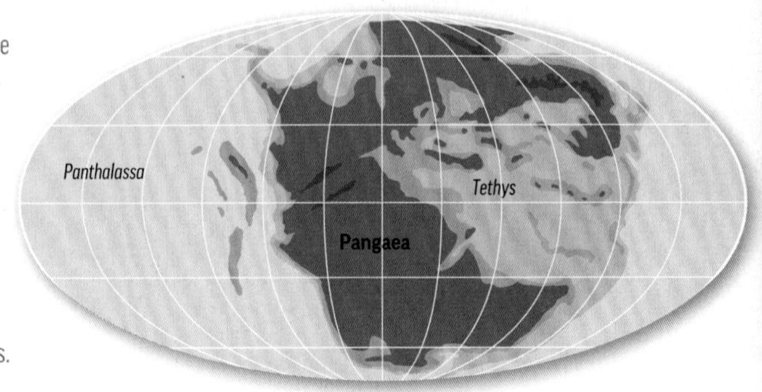

► Fig. 1.3. Earth's geography during the Triassic period. On the following page, **above**, a Late Triassic ecosystem, Morocco. A ferocious *Arganasuchus*, a rauisuchian, threatens a *Moghreberia*, a dicynodont ("mammalian reptile"). On the riverbank is an *Aetosaurus*, a small, armored herbivore. In the water, a *Paleorhinus*, a piscivorous phytosaur resembling a crocodile, observes the scene. **Below**, a Late Triassic ecosystem, Germany. A *Plateosaurus*, a herbivorous prosauropod, is feeding next to two *Procompsognathus*, small carnivorous theropods, and a *Megazostrodon* (lower left), an ancestor of the mammals.

both are, in turn, encircled by a giant ocean called Panthalassa. Over the course of the Triassic, plate tectonics leads to the gradual breakup of Pangaea. A rift (fracture of the oceanic crust) from east to west progressively opens between what will become North America and Eurasia, on the one hand, and Africa and South America, on the other. The result, during the Late Triassic, is a sundering of the supercontinent into a northern landmass (called Laurasia) and a southern one (Gondwana), as well as an expansion of the Tethys from east to west.

During the Early Jurassic, Earth's geography is shaped largely by the breakup of Gondwana into the West Gondwana plate (consisting of Arabia, Africa, and South America) and the East Gondwana plate (Madagascar, India, Australia, and Antarctica). Laurasia remains a continuous landmass, from the west coast of the United States all the way to Borneo. Western Europe lies on its southeastern coast. During the Triassic and the Early Jurassic, the central section of the Atlantic Ocean begins to open up (this process is called **oceanization**; fig. 1.4).

Later, during the Late Jurassic, tectonic plate collisions lead to the formation of the Andes in South America and the Sierra Nevada in North America. The Tethys attains its maximum dimensions, with an active east-west **oceanic ridge** from Australia to the Caribbean. Following Africa's

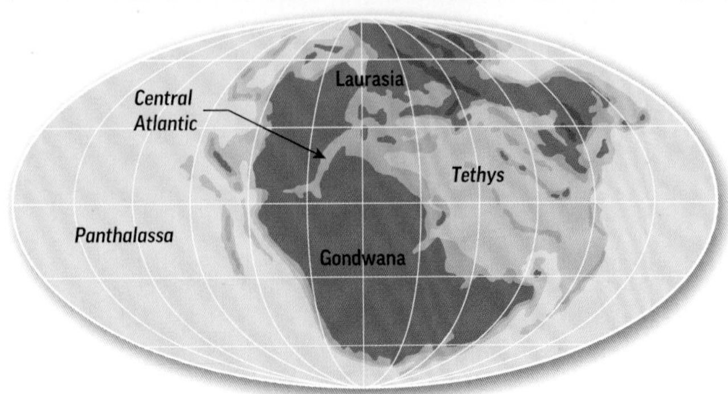

◄ Fig. 1.4. Earth's geography during the Early Jurassic. **Above**, a scene of Early Jurassic life, Antarctica. Two male *Cryolophosaurus* (carnivorous theropods 6–7 meters in length) fight over territory.

northward journey and counterclockwise rotation, the Tethys begins to close. The young Atlantic Ocean widens and starts to extend to the north, between Eurasia and North America (fig. 1.5).

During the Early Cretaceous, India splits from the rest of East Gondwana and begins to drift to the north (fig. 1.6). An interior ocean (the Western Interior Seaway) gradually takes shape in North America. The North Atlantic keeps opening up and, by end of the Cretaceous, separates Eurasia from North America, while the rapid expansion of the South Atlantic separates South America from Africa and Arabia (fig. 1.7). The Tethys, meanwhile, is caught in a double bind: to the south, the continental plates of Africa, India, and Australia, drifting northward, plunge beneath its oceanic plate; to the north, the Tethys itself is plunging beneath Eurasia. It is this dual drift of the plates beneath one another (the phenomenon of **subduction**) that leads to the gradual closing of the Tethys. The Alps and the Pyrenees are formed in the course of these tectonic clashes. The western Tethys starts to close, and the shared area between the Tethys and the Atlantic disappears, while India draws closer to Asia.

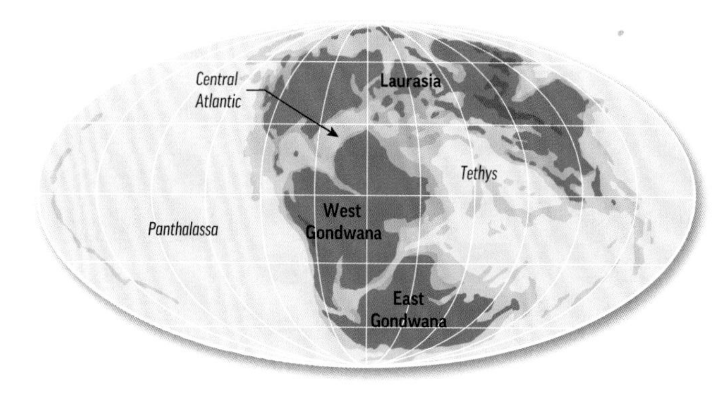

▲ Fig. 1.5. Earth's geography during the Late Jurassic. **Above**, a reconstruction of the Late Jurassic "Pterosaur Beach," Crayssac, the Lot valley (France). Thanks to the numerous footprints of pterosaurs (flying reptiles), theropods, and crocodiles at this celebrated site, scientists can reconstruct this coastal environment very precisely.

During the Cenozoic (66 Ma–present), the Tethys closes almost completely, and India collides with the Eurasian plate, leading to the formation of the Himalayas (fig. 1.8). The exact time of this event is the subject of debate because, although the *geological* data for the most part point to maximum isolation of India in the Late Cretaceous and to a collision with Eurasia around 55 million years ago, the *paleontological* data suggest a collision before the end of the Cretaceous (that is, earlier than 66 Ma). Around the same time, a passage toward the Arctic, between Greenland and Europe, takes shape.

► Fig. 1.6. Earth's geography during the Early Cretaceous. **Above**, an Early Cretaceous landscape, Europe. A couple of large brachiosaurids head for the trees, a group of *Iguanodon* leisurely cross their path, and a pterosaur alights. **Below**, an Early Cretaceous landscape, China. A *Yutyannus* (a tyrannosaurid) guards the corpse of a *Beipiaosaurus* (a theropod). On the left, a *Confuciusornis* flies off, while a diminutive *Jeholodens* (a small mammal) takes refuge in a tree on the right. Two pterosaurs fly over the lake.

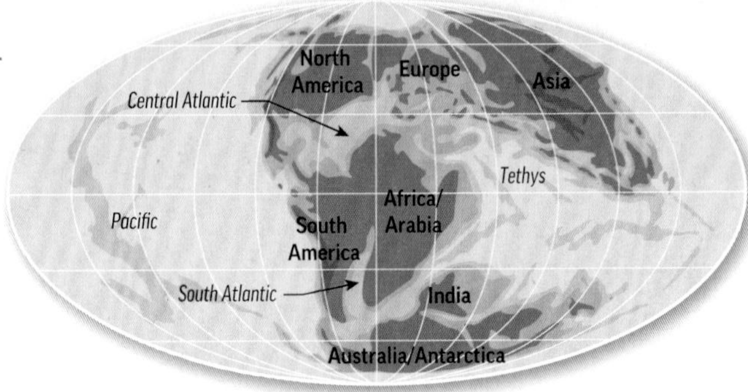

▶ Fig. 1.7. Earth's geography during the Late Cretaceous. **Above**, a Late Cretaceous landscape, Mongolia. An oviraptorid, *Citipati*, defends its nest from two dromaeosaurids, *Tsaagan*. Some ceratopsians, *Protoceratops*, can be seen in the far background, and a small enantiornith, *Gobipteryx*, takes refuge in a cycad on the upper left. **Below**, a Late Cretaceous landscape, North America. A *Triceratops* readies itself to charge a *Tyrannosaurus* to protect its offspring.

◄ Fig. 1.8. Earth's geography during the Cenozoic era. An Eocene (56–34 Ma) landscape, Messel (Germany): A *Gastornis*, a large terrestrial bird, surveys the shores of a lake next to two *Propalaeotherium*, forerunners of horses. A bird, *Messelornis*, feeds by the shore on the lower left, under the watchful gaze of an arboreal *Europolemur*, a lemur. A crocodile is on the lookout in the water, and two flying *Idiornis* complete the scene.

From One Crisis to the Next

The history of life on Earth is not one long, tranquil river. If evolution and the renewal of species in the course of time are a normal and continuing phenomenon, their thread is sometimes broken by prolonged periods of mass extinction, provoked by great changes in the environment, which reset the counters to zero. Some groups expire, while others take up the baton to continue the relay race into the future. The biological crises that arise during these periods sometimes have such a profound effect in the fossil record that geologists have used them to mark the subdivisions between geologic periods. Significant chronological cutoffs have therefore been placed at the level of the biological crises that carry their name today: the Permian/Triassic, the Triassic/Jurassic, and the Cretaceous/Paleogene crises.

A mass extinction has a three-part definition: it entails (1) a very significant drop in biodiversity (at least 75% of species disappear) (2) over the course of a relatively brief geologic time span (several million years at most) (3) on a global scale. In the early 1980s, American paleontologists Jack Sepkoski and David Raup identified five important mass extinctions among marine invertebrates: the Ordovician/Silurian extinction, around 445 million years ago (an 85% loss of species); the Late Devonian extinction, between 380 and 360 million years ago (75% loss); the Permian/Triassic extinction, around 252 million years ago (95%); the Triassic/Jurassic extinction, about 201 million years ago (75%); and the Cretaceous/Paleogene extinction, 66 million years ago (75%). Given the magnitude of these losses, other types of creatures at the time, both in water and (starting in the Devonian period) on land, must have been decimated as well. And we can certainly add the Holocene extinction, which began 10,000 years ago and is ongoing as a result of human activity. This extinction event is absolutely unique in the history of life on Earth, since the others were caused by "external" phenomena—whether temperature changes, sea-level fluctuations, volcanic eruptions, or meteorite impacts.

The beginning of the Mesozoic was characterized by the Permian/Triassic crisis, the planetwide disaster at the end of the Permian period (the last period of the previous era, the Paleozoic). It represents the most significant crisis that life on Earth was ever subjected to: roughly 95% of marine invertebrates vanished (compared with 75% during the famous Cretaceous/Paleogene crisis). The suspected principal cause of the disaster is tremendous volcanic activity in the region known today as Siberia. This was a kind of mega-eruption that continued for almost 600,000 years and that saw the outpouring of immense effusions of lava (or "magmatic provinces"), which solidified to form rocky structures known as **traps**. Even though they are less well known than the Deccan Traps (in India) that mark the Cretaceous/Paleogene boundary (see chapter 6, p. 179), the Siberian Traps are a layer of hardened lava covering millions of square miles, several times the size of the Deccan Traps.

The Siberian Traps are a layer of hardened lava covering millions of square miles, several times the size of the Deccan Traps!

This massive, long-lived volcanic activity spewed large quantities of carbon dioxide (CO_2) and toxic gases into the atmosphere, leading to a drastic rise in temperatures (between 5°C and 10°C) over several thousand years. This brutal warming unleashed rains that caused the intense leaching of continents, washing gigantic quantities of nutrients into the oceans and thus fueling a massive proliferation of algae and bacteria. (Evidence of this is found in the omnipresence of black sediments, rich in organic matter, above the Permian/Triassic stratigraphic boundary.) More algae and bacteria growing meant more algae and bacteria dying, and the decomposition of much greater than usual amounts of these organisms depleted the oxygen that was dissolved in the oceans, creating an environmental condition known as **anoxia**, which doomed numerous groups of other organisms living on the ocean floor. Elsewhere, the increase in atmospheric concentrations of carbon dioxide led to the acidification of the oceans and consequently affected the ability of certain marine animals, such as table corals and numerous mollusks, to form the calcium carbonate (also known as calcite; $CaCO_3$) needed for their skeletons. The effects of volcanic activity were amplified by the fact that, at the time, the continental masses were grouped together in a single mega-continent.

The climate deteriorated, which slowed oceanic circulation, leading to poor oxygenation of the ocean basins. In addition, a **marine regression** (drop in sea level) of roughly 250 meters led to significant draining of continental shelves and therefore reduction of the space available to marine life on these not-very-deep ocean floors (**benthic** life).

Even so, ancient Triassic environments were not significantly different from those of the Permian. The overall climate was warm, there were no polar ice caps, and the difference in temperature between the equator and the poles was less than it has been at other times in our past. Moreover, following the Permian/Triassic crisis the fauna and flora were considerably modified, and new groups, along with some survivors, filled the vacant ecological niches. This is how, starting at the beginning of the Triassic, the two principal groups of marine reptiles—ichthyosaurs and sauropterygians—diversified. The Middle Triassic was marked by an important marine regression. This epoch of low sea levels was followed by a **marine transgression** (a rise in sea level) that lasted until the end of the Triassic, one result of which was that Europe was almost completely submerged, with only a few mountain massifs—the Armorican Massif (northwestern France), the Ardennes (Belgium), the Massif Central (south central France), and some others—left above water, forming archipelagos. A new crisis arose at the end of the Triassic (see chapter 4, p. 134), but its causes are still poorly understood. Once again, volcanic activity is the principal suspect. It was around this time that a gigantic magmatic province developed on the spreading Central Atlantic seafloor. This geologic hotbed would have, in a manner similar to that of the Permian/Triassic crisis, emitted large quantities of carbon dioxide and toxic gases, leading to global warming, ocean acidification, and areas of anoxia in the oceans.

Temperatures may have reached record highs during the Early Jurassic, when Earth's polar regions appear to have been home to tropical vegetation.

In the Jurassic (201–145 Ma), the groups of animals that survived the Triassic and the new ones that evolved encountered a world vastly different from the preceding one. This was the point at which the plesiosaurs, which had arisen at the very end of the Late Triassic, diversified. For a long time, scientists believed the Jurassic to have been a period of great climatic stability, but the climate was globally both warmer and more humid than it had been during the Triassic. Indeed, there is evidence of record temperatures during the Toarcian age, around 180 million years ago; it appears that Earth's poles were home to tropical vegetation (although very little data is available for the South Pole, since it is currently covered in ice). Other ages of the Jurassic period were less warm, though temperatures were still higher than those of today. Probably because of changes in oceanic currents, tied to the opening of a vast area of communication between the Pacific and the Tethys, by the end of the Jurassic the arid climate of the Middle Jurassic had given way to a warmer and more humid one for the continents surrounding the Tethys. The climatic changes and the rise in temperatures enabled corals to create gigantic reefs.

At the beginning of the Cretaceous, those lands bordering the Central Atlantic were still under the influence of a humid climate. However, during the Late Cretaceous, a large marine transgression, which peaked at the end of the Cenomanian age (ca. 95 Ma), partially submerged all the continents: sea level rose to nearly 250 meters higher than today! New **epicontinental** seas (situated above continental, not oceanic, plates) formed, most notably in the Saharan regions of Africa, where the Trans-Saharan Seaway made its appearance. In Europe, during the Late Cretaceous, the accumulated skeletons of marine microorganisms in "the Chalk Sea" formed chalk deposits, notably in the Parisian basin, that in early modern times were recognized as proving the existence of this ancient sea (see chapter 6, p. 164). To the south it was phosphates, instead of chalk, that were deposited in massive quantities along an immense latitudinal band running from the Middle East across North and West Africa all the way to eastern Brazil (see chapter 6, p. 172).

At the end of the Cretaceous period, the Arctic Ocean started communicating with the Atlantic. As a result, thermohaline circulation (also known as the ocean conveyor belt), which regulates the distribution of heat at the planet's surface, was altered, and temperatures fell globally. The division between the Cretaceous and the Paleogene (the first period of the Cenozoic era) is marked by the most famous of the mass extinctions, most notably because it saw the demise of all non-avian dinosaurs (birds, which are dinosaurs, survived!). This crisis has been very well studied, but its causes are still debated (see chapter 6, p. 179).

Life in All Its States

During the Mesozoic, as during all other periods of the history of life, many species saw the light of day only to shortly disappear. Dinosaurs first appeared in the Triassic, around 230 million years ago; among these dinosaurs, birds (e.g., *Archaeopteryx*) arose in the Late Jurassic, 146 million years ago. The mammals from which we are descended made their debut in the Early Jurassic (about 200 Ma), and the angiosperms (flowering plants) made theirs during the Early Cretaceous.

Some Mesozoic animals were enormous. One was *Argentinosaurus*, a titanosaur sauropod (an immense herbivorous quadruped)

▼ Fig. 1.9. A size comparison of the largest fish, *Leedsichthys*, and a large pliosaur, *Liopleurodon*.

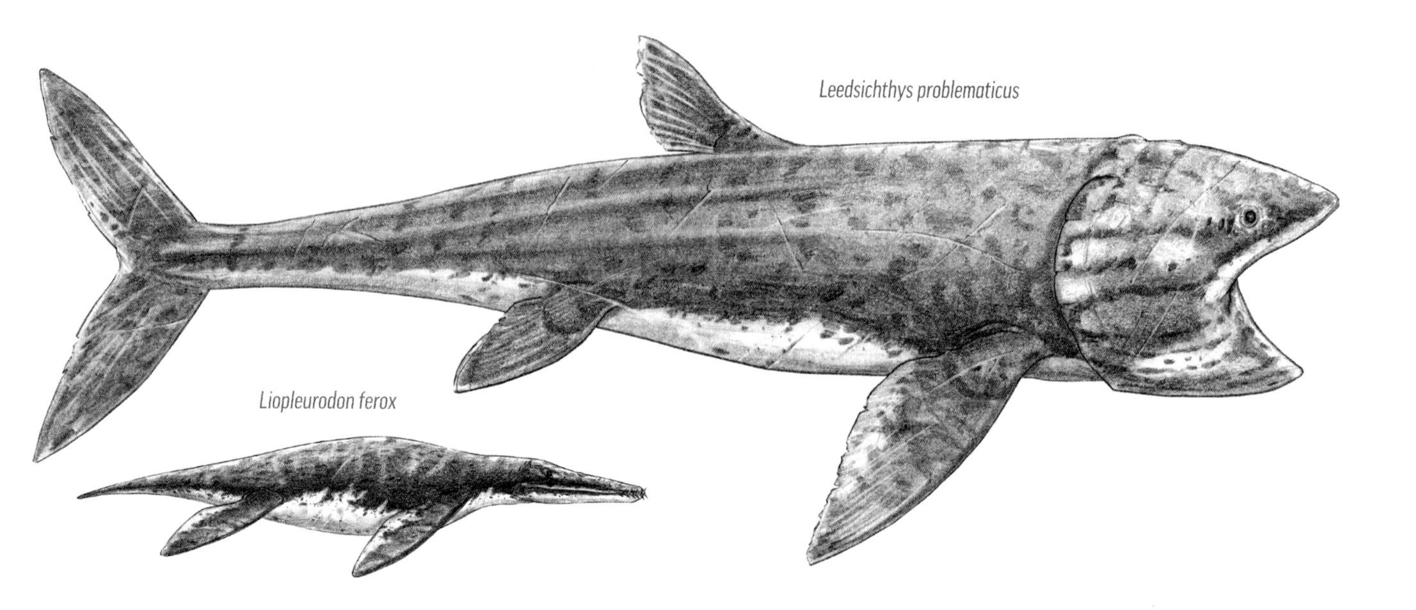

▲ Fig. 1.10. A Late Cretaceous lamniform (mackerel) shark, *Cretoxyrhina*, attacking a mosasaur, *Clidastes*.

from the Cretaceous in Argentina, which attained a length of almost 35 meters. But another giant—taking relative proportions into consideration—was *Clatrotitan*, a Triassic insect belonging to Titanoptera (an extinct order related to modern crickets), which, as the name suggests, was titanic: its wingspan could exceed 40 centimeters! We shall not proceed any further in our review of the numerous groups of land-dwellers during the Mesozoic. We just felt we should mention some notable types of animals that figuratively rubbed shoulders, or wings, with the marine reptiles.

The warm seas of the Mesozoic abounded in life: crinoids (sessile or mobile echinoderms with both a spindle and arms, thus resembling plants; fig. 4.10, pp. 124–25), corals, cephalopod mollusks (a group today represented by nautiluses, octopuses, and squids), cartilaginous fishes (class **Chondrichthyes**), bony fishes (class **Osteichthyes**), and more. These species, which generally belonged to the lower levels of the food chain, were, just as in today's oceans, much more numerous and abundant than marine reptiles, which were above them in the food chain. It was during this period that many of the modern groups of cartilaginous fishes and ray-finned fishes (subclass Actinopterygii, in Osteichthyes) first appeared and began to diversify. Pachycormid fish, for example, also called "whale-fish," which were **actinopterygian** filter feeders, were among the largest fish ever. The record holder, *Leedsichthys* (fig. 1.9) exceeded 20 meters in length. In comparison, today it is among the cartilaginous fishes that we find the largest fish: the whale shark (*Rhincodon typus*), a peaceful eater of plankton, can reach a length of 20 meters and weigh up to 34 tons. The biggest modern bony fish is the giant oarfish (*Regalecus glesne*), a ribbon-shaped creature that can reach 11 meters in length and is probably the origin of the many tales about sea serpents.

The class Chondrichthyes includes sharks and rays. Among the most emblematic sharks of the Mesozoic were the hybodonts, which had two dorsal (topside) fin spines (a typical feature of the group) and could reach 3 meters in length. Sharks were among those very rare animals that could be predators of marine reptiles (fig. 1.10). Some sharks ate the remains of marine reptiles; however, traces of scars on marine reptiles' bones show that sharks preyed on them even while they were alive. The reverse is also the case: certain sharks were the prey of large marine reptiles, such as mosasaurs.

Among the Mesozoic invertebrates and potential prey of the marine reptiles, the most famous are without a doubt the ammonoids (ammonites and *Ceratites*), cephalopod mollusks with rolled-up shells varying in diameter from several millimeters to 2 meters (fig. 1.11). *Ceratites* first appeared in the Permian period, ammonites followed in the Jurassic period, and ammonoids reached their apex before disappearing at the very end of the Cretaceous. Extremely varied and fast evolving, these animals are very useful for geologists because their fossils are used to subdivide the various layers of sedimentary rock and therefore to calibrate geologic time. So, each period of the Mesozoic and each division within those periods is characterized by the presence of a species or a group of species—in other words, specific ammonoids—in its sediments: this is what is called a **biozone**. Biozones can span millions of years.

Mesozoic seas were filled with other famous types of cephalopods as well, known as belemnoids (belemnites). These animals, with a pointed internal shell (called a phragmocone), first appeared in the Carboniferous period of the Paleozoic era and can be regularly found in Mesozoic deposits, next to the shells of ammonites and the remains of marine reptiles (fig. 1.12). All the creatures above constituted some of the numerous sources of food for marine reptiles, but in some instances they were predators of marine reptiles too. Their shapes and their variety, as well as those of the marine reptiles, often changed during the Mesozoic, as both prey and predators were constantly evolving.

▲ Fig. 1.11. *Hildoceras*, an Early Jurassic ammonite typical of the Toarcian age.

▼ Fig. 1.12. *Cryptoclidus*, a Middle Jurassic plesiosaur, chases a school of belemnites.

CHAPTER 2

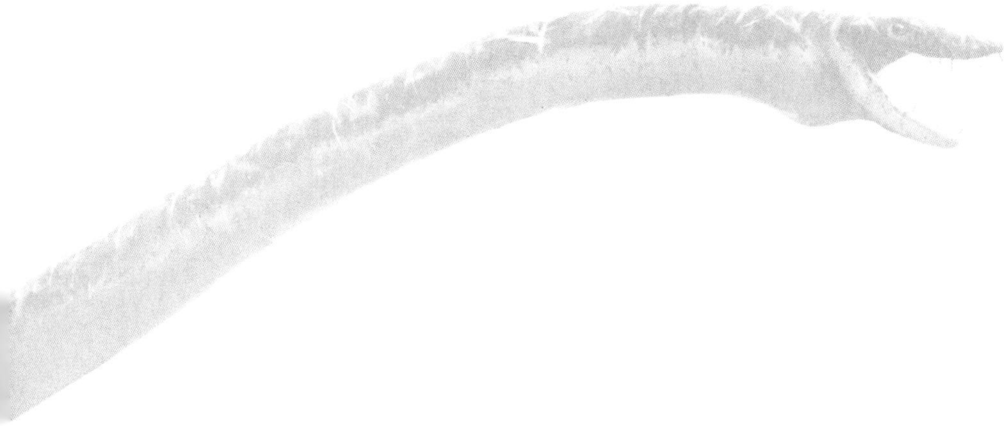

ALL AQUATIC, ALL DIFFERENT

> We have now reached those of all the reptiles, and maybe of all the fossil animals, which least resemble those we know, and which seem to be made to most surprise the naturalist by way of combinations of structures which, without a doubt, will arouse incredulity in those who are not able to observe them in person. In the first case the muzzle of a dolphin, the teeth of crocodiles, the head and the sternum of a lizard, the flippers of cetaceans, four in number, and the vertebrae of fish; in the second case, together with these same cetacean flippers, the head of a lizard with a long neck resembling that of a lizard. ... The Plesiosaurus is perhaps the most bizarre of the inhabitants of the ancient world, as well as being the one that, more than all others, seems to deserve the qualification of monster.
>
> —Georges Cuvier
> *(Recherche sur les ossemens fossiles,* 1836, p. 387 [translated]).

As we emphasized previously, the vast majority of modern reptiles are terrestrial. This is also true of extinct reptiles, especially the most ancient ones. To understand why, let us examine what the word "reptile" means more closely.

I

Reptiles? What Are Those?

Reptiles are vertebrates with four legs. Expressed more scientifically, they are **tetrapods**: they have four limbs with fingers, just as amphibians (frogs, salamanders, etc.) do. However, reptiles are **amniotes**: a reptile embryo develops inside an egg filled with fluid contained by a membrane (the amnion). This amniotic fluid functions as a substitute for an aquatic environment and protects the embryo from dehydration (fig. 2.1). The result is that reptiles, unlike their amphibian ancestors, do not need to lay their eggs in water. A major evolutionary innovation, the amniotic egg allowed reptiles to colonize new environments, on dry land.

Amniotes emerged during the Late Carboniferous, about 310 million years ago. Unfortunately for paleontologists, no amniotic eggs from that time have been discovered. Moreover, these first amniotic eggs most likely did not possess any protective calcareous (calcium-containing) shells—which must have appeared later in the evolutionary process—but were probably enveloped in a flexible

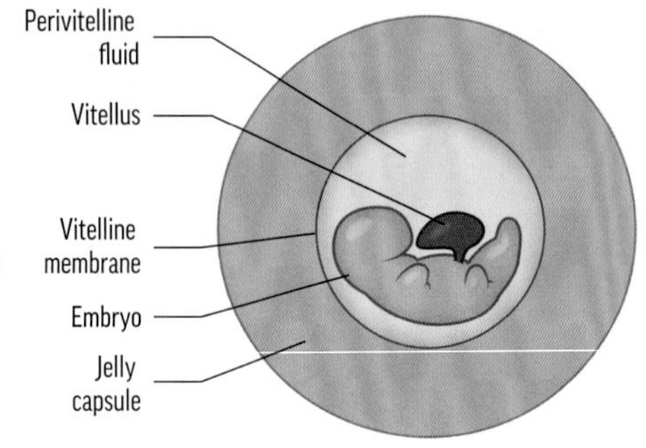

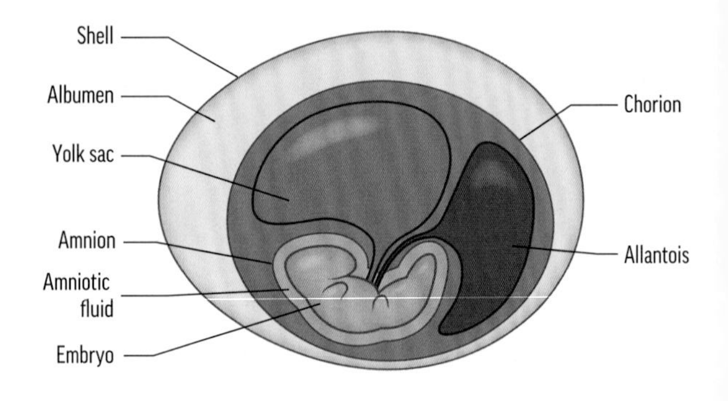

► Fig. 2.1. A comparison of a non-amniotic egg (from a fish) with an amniotic egg (from a bird).

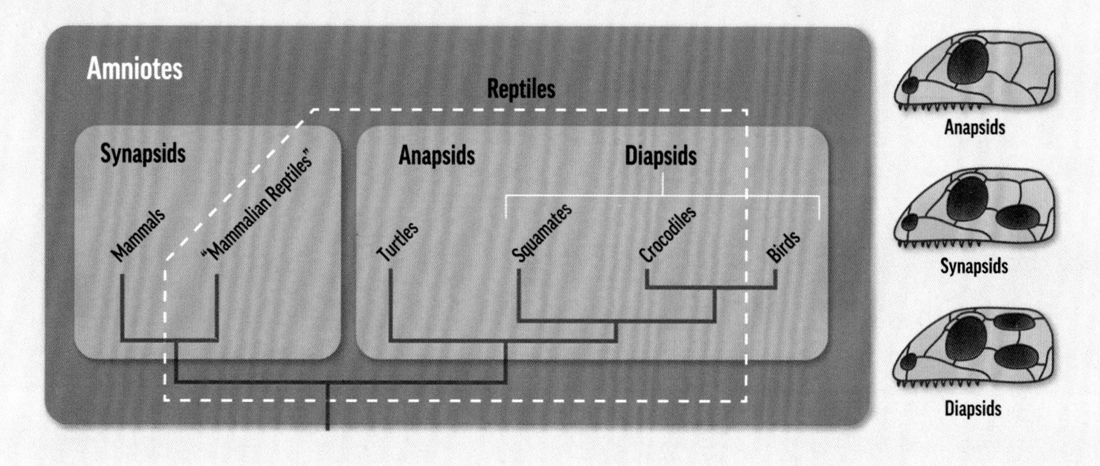

◄ Fig. 2.2. The classification of amniotes, as a function of their temporal fossae. A dotted line surrounds the groups that were originally classified as reptiles (see "Reptiles? Sauropsids? What's the Fuss About?," p. 27).

membrane, as leatherback turtles' eggs are. So how do we know that these Late Carboniferous tetrapods were amniotes? How can we tell an amniote from a non-amniote without seeing its eggs? Well, today's amniotes all exhibit signs of adaptation to life on dry land—for instance, a **keratinized** skin that limits cutaneous evaporation and thus keeps the animal from becoming dehydrated by exposure to open air. They also have more efficient lungs, thanks to a pleated structure that increases the surface area for gas exchange. However, many of these characteristics are tied to soft body parts that fossilize only in extremely rare conditions. Paleontologists must therefore be guided by telltale skeletal features of amniotes. These include the absence of an otic notch (a crease in the posterior margin of the skull roof, one behind each eye socket) and the presence of a single occipital condyle (a rounded knob of the occipital bone, on the rear of the skull, that allows for its articulation, or joining, with the spine) rather than two.

During the Late Carboniferous, one group of amniotes began to differentiate themselves from the others. These were the synapsids. They can be distinguished from reptiles (or sauropsids—see "Reptiles? Sauropsids? What's the Fuss About?," p. 27) by certain differences in their skulls (fig. 2.2). In a reptile's skull, there is a pair of openings over the palate (the bone that forms the roof of the mouth); these openings are called the infraorbital foramina. There also may be openings on each side at the rear of the skull, called temporal cranial fossae. Synapsids, on the other hand, have a single opening in a lower position on each side of the skull, bordered by the squamosal, postorbital, and jugal bones. It is from the synapsids that mammals would later emerge, during the Jurassic period. The synapsids of the Carboniferous included the **paraphyletic** group once classified as "mammal-like" or "mammalian reptiles": they were in fact nonmammalian synapsids.

▼ Fig. 2.3. A simplified classification of reptiles, modern and extinct, with their principal groups (turtles, archosauromorphs, and lepidosauromorphs).

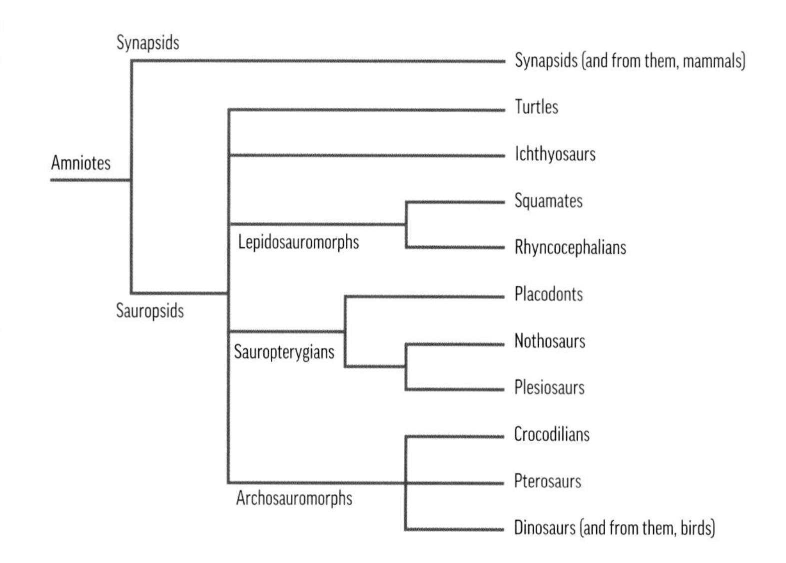

Among reptiles, anapsids are characterized by the absence of temporal fossae. They include several groups from the Permian, such as the extinct pareiasaurids (which were massive, stocky, and armor-clad. The **diapsids** exhibit two pairs of temporal fossae, although these may be greatly modified (e.g., by the loss of the lower opening). They include numerous groups, the three most important being the ichthyosaurs, the **lepidosauromorphs**, and the **archosauromorphs** (fig. 2.3). The lepidosauromorphs include the squamates, which have a long body and are covered with horny scales. Lizards, snakes, and the extinct mosasaurs belong to this order (Squamata). Lepidosauromorphs also include sauropterygians (placodonts, nothosaurs, plesiosaurs, etc.), now extinct. The archosauromorphs include the prolacertiforms (reptiles that looked like large lizards), to which the famous *Tanystropheus* (see fig. 4.4, pp. 116–17), as well as the archosaurs, belong. The archosaurs include both living and extinct crocodiles, pterosaurs, and the famous dinosaurs ... birds included. Birds are therefore reptiles!

Turtles are also reptiles, but their classification is problematic (we shall return to that issue). For a long time, they were considered anapsids because of their lack of temporal fossae; today, many experts think they are diapsids in which the temporal fossae closed. According to different **phylogenetic** (evolutionary) analyses (i.e., morphological vs. molecular), turtles are identified as either lepidosauromorphs or archosauromorphs. The phylogenetic position of the mesosaurs (see chapter 3, p. 103), some of the earliest marine reptiles, also remains problematic.

Reptiles Become Very Small, Biding Their Time

Quite independently of one another, during the Permian period and the subsequent Mesozoic era, many groups of reptiles—the mesosaurs, the ichthyosaurs, and the sauropterygians, as well as several groups of turtles, squamates, and crocodylomorphs—colonized the marine environment. What were the circumstances that accompanied these numerous "returns to the water"? At this point in our story, let us go back a little way and observe what was occurring on land at the end of the Paleozoic.

The Permian proved an especially prolific time for amniotes: they exploded in diversity, thanks to ferocious competition between synapsids and reptiles. It was during this period that reptiles became more "discreet"—in other words, more modestly sized, usually no longer than half a meter. Ecosystems were dominated by large synapsids, both carnivorous ones and herbivorous ones. *Dimetrodon* (fig. 2.5) is the best known among them. Pelycosauridae, the group to which *Dimetrodon* belongs, were about 3 meters in length and preyed on other Permian synapsids, such as dicynodonts (see fig. 1.3, p. 11), a very diverse group of herbivores ranging from about half a meter to several meters in length. Yet reptiles occupied practically all the terrestrial ecological niches and exhibited a wide variety of adaptations. It was from within this group that the first bipeds emerged, displaying slender and agile forms, but it was also at this time that the first arboreal (tree-dwelling) and gliding vertebrates appeared. In addition, amniotes made their first notable return to the aquatic environment in the Permian. This adaptation occurred among the reptiles—specifically, the mesosaurs. A sketch of the forces on the ground at the end of the Permian is

Reptiles? Sauropsids? What's the Fuss About?

For most people, reptiles seem to be well defined: they are tetrapods that normally crawl, lay eggs, and are covered in scales; their body temperature is variable (they are **ectotherms**)—in short, they are the complete opposite of mammals ... and this is the problem! Following is a short overview of the changes the definition of *reptile* has undergone over the course of more than two centuries.

The classification "reptile" was created by Swedish naturalist Carl Linnaeus in 1758, who placed reptiles among the amphibians. Several decades later (in 1824), French anatomist and paleontologist Georges Cuvier included the batrachians (anurans and urodeles) among the reptiles. Cuvier therefore considered all vertebrates that are devoid of feathers, hair, and breasts; that breathe air in their adult stages; and that move by crawling—in other words, all modern amphibians and all reptiles in the traditional sense (excluding birds)—as belonging together. In 1866 German biologist, artist, and philosopher Ernst Haeckel regrouped reptiles, birds, and mammals together within the category of Amniota. At about the same time, British biologist and paleontologist Thomas Henry Huxley proposed excluding amphibians from the reptiles and dividing vertebrates into three groups: Sauropsida (reptiles and birds), Ichthyoidea (fish and amphibians), and Mammalia (mammals).

Later the classification Sauropsida fell out of use and the classification Reptilia, devoid of a true definition, ended up encompassing the "stem group" of reptiles, birds, and modern mammals ... in other words, with some approximation, the amniotes! Nevertheless, the notion of "reptile," which in people's minds was opposed to the notion of "mammal," continued to persist and, at the beginning of the twentieth century, some strange fossils led to the concept of "mammalian reptile." This concept allowed the grouping of several types of fossil animals that anatomically resembled mammals but retained some reptilian characteristics (e.g., *Dimetrodon* and *Dicynodon*) (fig. 2.5) under the same name. In 1903, American paleontologist Henry Fairfield Osborn regrouped the set of reptiles that exhibit a single temporal fossa, or a single opening on either side of the skull (thereby excluding crocodiles, turtles, and lizards), under the heading Synapsida.

The division of the amniotes into mammals, birds, and reptiles persisted into the 1970s, when the introduction of a new system of classification, **cladistics**, underscored the kinship relations among groups (i.e., **phylogeny**) and changed experts' perspectives on the situation. In the older systems, birds were excluded from the reptiles, while their dinosaur ancestors were included (which led to reptiles being a paraphyletic group, one that includes a common ancestor but only some of its descendants). Thanks to this method, the label "sauropsid" made a splashy return to the scene! Amniotes were then divided into two groups: (1) sauropsids, comprising turtles, archosaurs, and lepidosaurs (i.e., all the descendants of the common ancestor, and all modern reptiles, meaning birds too), as well as some extinct groups without current representatives; and (2) synapsids, comprising both mammals and the ancient "mammalian reptiles," which are today deemed nonmammalian synapsids. Some problematic fossil amniotes, the mesosaurs (which we shall return to; see chapter 3, p. 103), were then considered to be nonreptilian sauropsids, but their classification has recently changed, and today they are considered reptiles. In this case the notions of "sauropsids" and "reptiles" are equivalent (the terms are synonyms). This is the reason why today a considerable number of researchers once again make use of the term "reptiles" when dealing with matters of phylogeny, and they define reptiles as the set of amniotes that are closer to lizards and crocodiles than to humans. Expressed somewhat differently, reptiles are the **monophyletic** group comprising turtles, squamates, crocodiles, and birds, as well as their common ancestor and all their descendants.

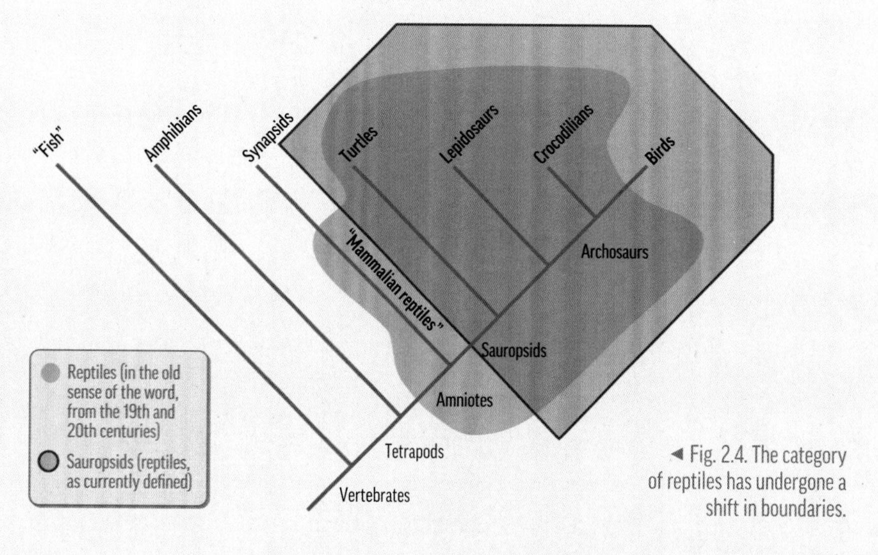

◄ Fig. 2.4. The category of reptiles has undergone a shift in boundaries.

▲ Fig. 2.5. *Dimetrodon*, a Permian synapsid, devouring a temnospondyl amphibian.

roughly as follows: large synapsids dominated the terrestrial ecosystems, while reptiles, smaller and more discreet, occupied a wide variety of ecological niches. But the winds of change were blowing, and soon they would favor the reptiles.

Long Live the Crisis!

The end of the Permian, and therefore the end of the Paleozoic, about 252 million years ago, was characterized by a radical change in the biological landscape: almost 95% of all species in the oceans disappeared. On land, the pelycosaurs did not survive the crisis, and the large herbivorous synapsids were also greatly affected. The placoderms (a class of armored fish with powerful jaws) (fig. 2.6, right) and the **trilobites** (a group of marine arthropods) died out (fig. 2.6, left). Other groups came very close to extinction, such as the foraminifera (an order of tiny animals with shells; 97% of species were lost), the bryozoans (a phylum of sessile animals that outwardly resemble coral; 79% were lost), **brachiopods** (a phylum of marine invertebrate; 96% were lost), echinoderms (which include both sea urchins and starfish; 98% were lost) (fig. 2.6, center), gastropods (a class of mollusks; 98% were lost), and the ammonoids (a subclass of cephalopods; 97% were lost). The sharks were also severely affected. What was the impact of the Permian/Triassic crisis on reptiles? Data are, unfortunately, not yet complete enough to allow precise analysis.

A new "post-crisis" world was thus outlined at the beginning of the Triassic (the first period of the Mesozoic), characterized by a wide-ranging renewal of the fauna. Although synapsids were still present—for instance,

there were the herbivorous dicynodonts—they were reduced in size. Mammals first appeared during the Late Triassic or shortly after the Triassic/Jurassic crisis, but, unable to attain more than modest dimensions, they would remain in the reptiles' shadow until the end of the Cretaceous. The commanding heights of the food chains were gradually monopolized by the reptiles, which diversified in a spectacular manner, once again occupying a great many ecological niches and sometimes adopting improbable morphologies like those of the avicephalans (which lived in trees) or the pterosaurs.

▲ Fig. 2.6. Three marine victims of the Permian/Triassic crisis: the trilobites (a type of arthropod), the blastoids (a type of echinoderm), and placoderm fish.

Mesozoic Marine Reptiles Were Not Dinosaurs

Because of their "reptilian" aspect and their sometimes spectacular size, marine reptiles are often considered dinosaurs. Newspaper headlines have sometimes announced the spectacular discovery of a large "marine dinosaur." But this designation, used for purposes of simplification or to grab the reader's attention, is completely false. Would we describe a whale as a "pachyderm," on the basis that both whales and elephants are large mammals? Probably not.

Dinosaurs form a **monophyletic** group—that is to say, one that comprises the descendants of one common ancestor and is defined by the shared characteristics derived from this common ancestor. For instance, the upper end of every dinosaur's femur (upper leg bone) featured a well-defined neck and head. The dinosaurs were continental reptiles, as distant from "marine reptiles" as elephants are from whales. One species of dinosaur—*Spinosaurus aegypticus*—is thought to have been semiaquatic, but no dinosaur is considered to have been truly aquatic, not to mention marine.

Well, in that case, what are marine reptiles? Just like the designation "marine mammals," "marine reptiles" is devoid of phylogenetic value because it does not designate a group of species that share a unique and exclusive evolutionary origin. Its only purpose is simplification, to designate a widely varied group of reptiles that lived and fed in the same environment: the ocean.

"Marine reptiles," for purposes of classification, therefore form not a monophyletic group but a **polyphyletic** group, which contains a certain number of species that do not share a common ancestor. Just as marine mammals—which include unrelated

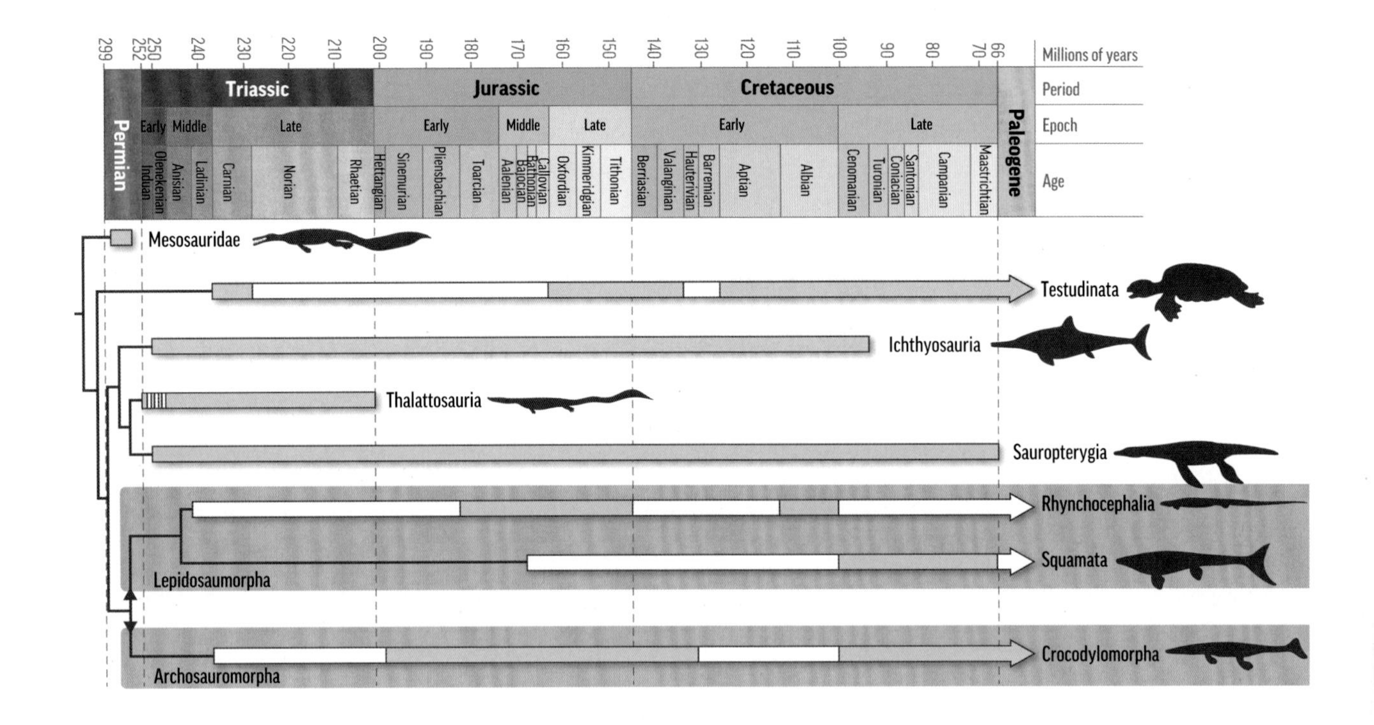

▲ Fig. 2.7. A general phylogeny, which includes all the groups to which marine reptiles discussed in this work belong. Blue indicates the presence and white indicates the absence of marine representatives at the time in each group.

lines, such as cetaceans (whales, dolphins, and porpoises), sirenians (dugongs and manatees), and pinnipeds (walruses, seals, and sea lions)—did much later, several groups of reptiles we shall now get to know better (fig. 2.7) "took to the water" in the Permian (286–249 Ma) and even earlier, and some were true pioneers (see chapter 3, p. 101). Although originally terrestrial, they invaded the Mesozoic oceans in a completely independent fashion. This secondary return to a life in water is considered a major evolutionary phenomenon in the history of the vertebrates, comparable to the tetrapods' "exit from the water" around 340 million years ago.

Reptiles with Large Eyes: The Ichthyosaurs

The word "ichthyosaur" dates from 1818 and derives from the Greek *ichthys*, "fish," and *sauros*, "lizard." Because they looked so different from living reptiles, these "lizard-fish" have intrigued paleontologists since their fossils were first discovered in 1699. They were not even properly recognized as reptiles until the nineteenth century.

The ichthyosaurs represent one of the most important groups of marine reptiles of the Mesozoic, as their widespread distribution and their extensive fossil record bear witness to. More than ninety species are known. They lived from the Early Triassic (around 252 Ma) until the beginning of the Late Cretaceous (roughly 90 Ma) (fig. 2.8). While most other

Early Triassic reptiles that invaded the marine environment stayed close to land, ichthyosaurs soon adapted to a pelagic way of life. Of all the reptiles in Earth's history, no other group has possessed a set of characteristics so extensively tailored to a life at sea. Among today's amniotes, only cetaceans are as suited to the open ocean as the ichthyosaurs were.

A Lot, a Little Bit, a Lot, More Than Anything ...

The ichthyosaurs first appeared in the early Triassic and diversified rapidly, taking advantage of the ecological niches left vacant by the numerous groups of sharks that had perished during the Permian/Triassic crisis. This hypothesis is supported by the ichthyosaurs' degree of diversity and their wide distribution during the Early Triassic, with specimens having been recovered in Canada, China, Thailand, and Japan, as well as on the Norwegian island of Svalbard. Since even the most primitive and ancient ichthyosaur fossils exhibit characteristics that suggest a high degree of adaptation to aquatic life, ichthyosaurs don't bear much resemblance to other groups of reptiles. As a result, paleontologists have been unable to determine their origin and from which group of Paleozoic terrestrial reptiles they descended.

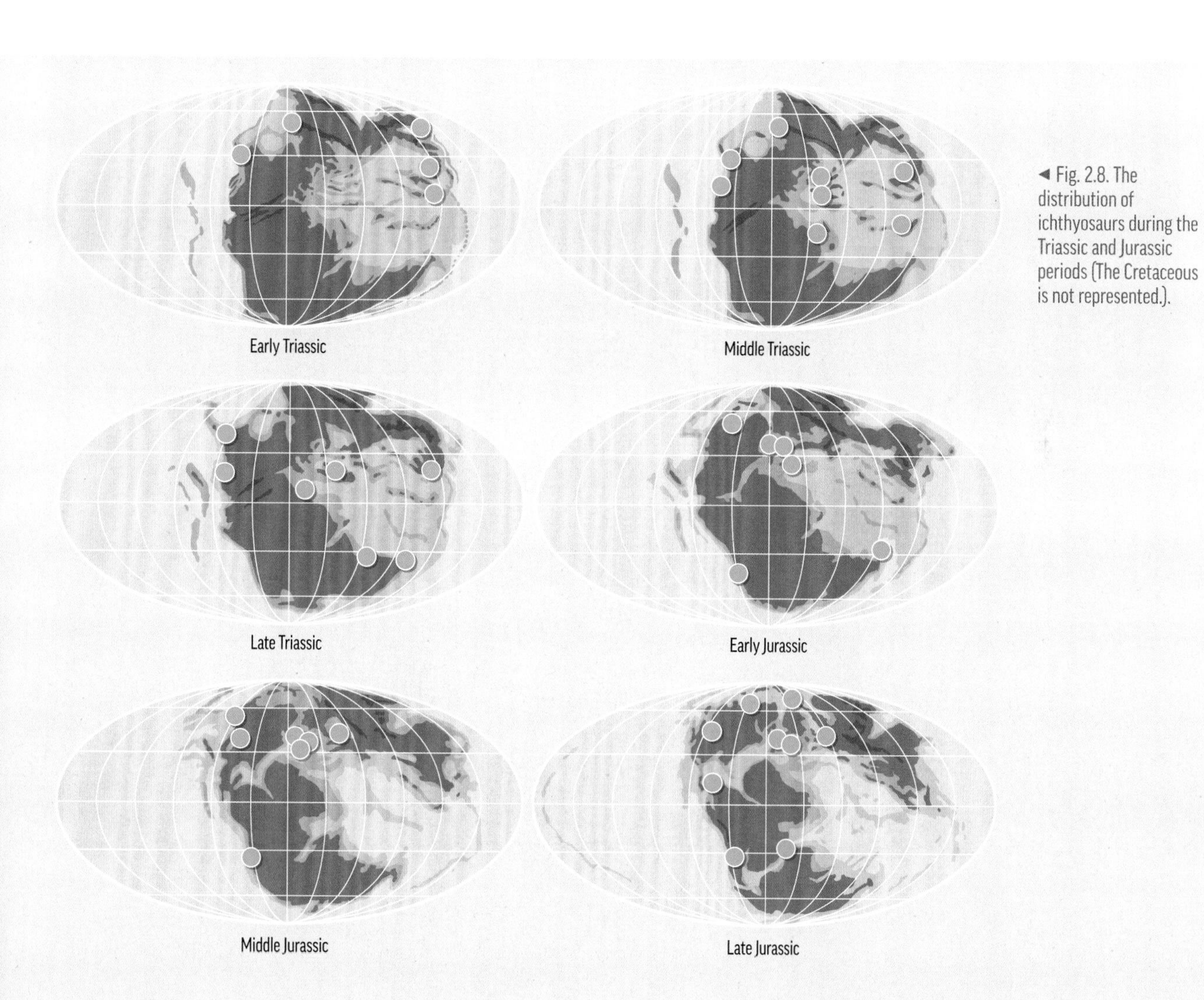

◄ Fig. 2.8. The distribution of ichthyosaurs during the Triassic and Jurassic periods (The Cretaceous is not represented.).

▲ Fig. 2.9. *Cartorhynchus*, an ichthyosaur from the Lower Triassic in China.

The 2014 discovery of *Cartorhyncus* (fig. 2.9), from the Lower Triassic in China, has, however, contributed some new information that is relevant to providing answers. This genus, known from only one very small specimen of an individual that must have been about 40 centimeters long, is seen as an ichthyosauriform—in other words, it has been placed in a lineage that leads to the ichthyosaurs proper. This individual was endowed with an elbow joint and was therefore, according to the paleontologists who studied it, still capable of using its limbs for terrestrial locomotion. Although that claim is not yet fully accepted, it is clear that this animal's adaptations to an aquatic lifestyle were more primitive than those of ichthyosaurs. Yet, as suggested by the **pachyostosis** (see "The Secrets of Bone," p. 129) at the level of its ribs, it seems to have been essentially, if not exclusively, aquatic, likely inhabiting shallow coastal waters.

Ichthyosaurs' range of specialization peaked during the Late Triassic, after which it gradually declined, except among the parvipelvians ("small-pelvised" ichthyosaurs), which diversified greatly during the Jurassic. The Triassic/Jurassic transition therefore corresponds to a time of major changes in the evolutionary history of the ichthyosaurs.

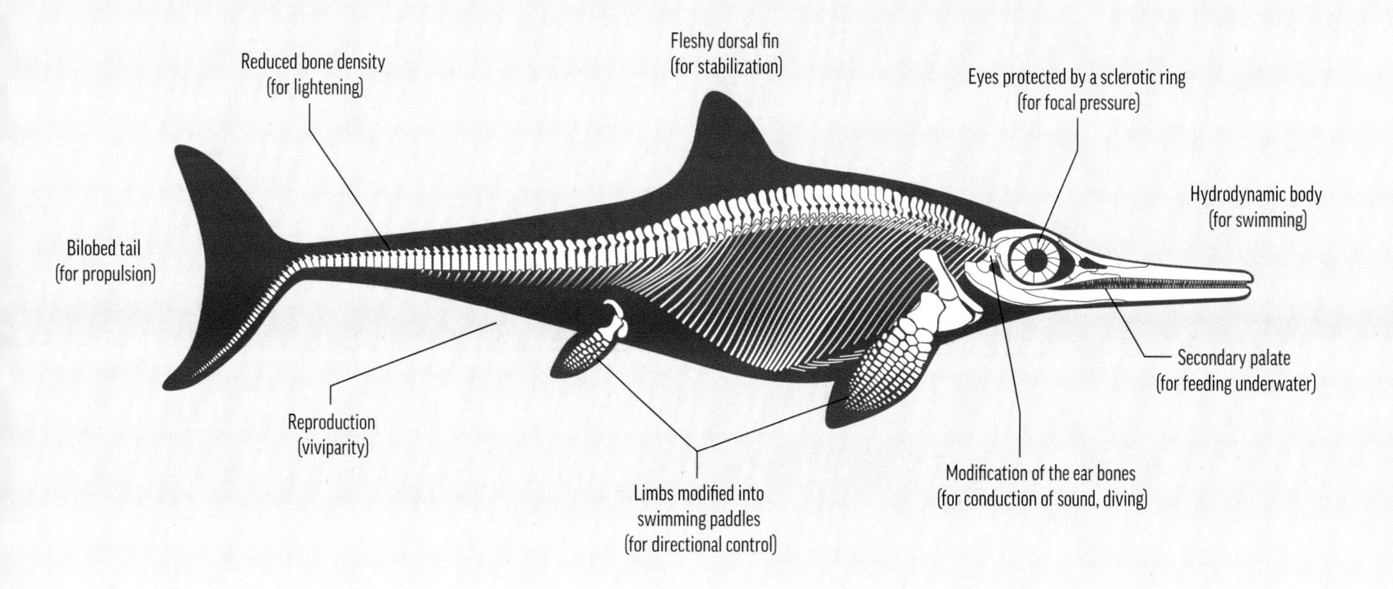

▲ Fig. 2.10. The principal anatomical and physiological characteristics of ichthyosaurs.

During the Jurassic, various lineages appeared and disappeared. Thanks to a family of ichthyosaurs called Ophthalmosauridae, the group enjoyed another golden age at the end of the Jurassic and in the Early Cretaceous. The end of the Cenomanian age, however, saw ichthyosaurs' rapid extinction, perhaps as a result of competition with other mega-predators: sharks were enjoying a real boom at the time, and other large marine reptiles, the mosasaurs, were beginning to diversify too. Until very recently it was believed that the ichthyosaurs of the Cretaceous were a largely similar bunch (represented by a single genus), but studies have shown that this was not the case. Instead, the ichthyosaurs' diversity—both systematic and ecological—and their geographic range remained significant until their brutal extinction about 90 million years ago.

But What Was So Special about Them?

Ichthyosaurs exhibit a single superior temporal fossa (see p. 26). They are nevertheless considered diapsids, since they evolved from diapsids, but their lower temporal fossa eventually disappeared. They were generally endowed with a fairly lengthy snout (in some forms, such as *Eurhinosaurus* (fig. 2.13) and *Excalibosaurus*, very long and sword-like); large eyes; nostrils high on the snout and close to the eyes; a short neck; a long, torpedo-shaped body; and both dorsal (top) and caudal (tail) fins (fig. 2.10). The shape of their vertebrae, which resemble those of sharks (i.e., discs with biconcave articular facets), is highly unusual among amniotes. In addition, they possessed a series of ventral (underside) bones, the gastralia, also referred to as ventral ribs. Unlike normal ribs, gastralia are not attached to the skeleton, and they form a series of transversal bones running from the sternum to the pubis.

Ichthyosaurs' limbs had undergone extreme transformation into swimming paddles (fig. 2.11): All the bones exhibit a simplified shape that tends toward rounded forms, without the significant ridges or edges that can be seen frequently in terrestrial reptiles. The bones were also flattened, and the joints between them were immobile, making the paddles very rigid. While the most ancient ichthyosaurs had five fingers (as their terrestrial ancestors did) and do not exhibit any hyperphalangy (an increase in the number of

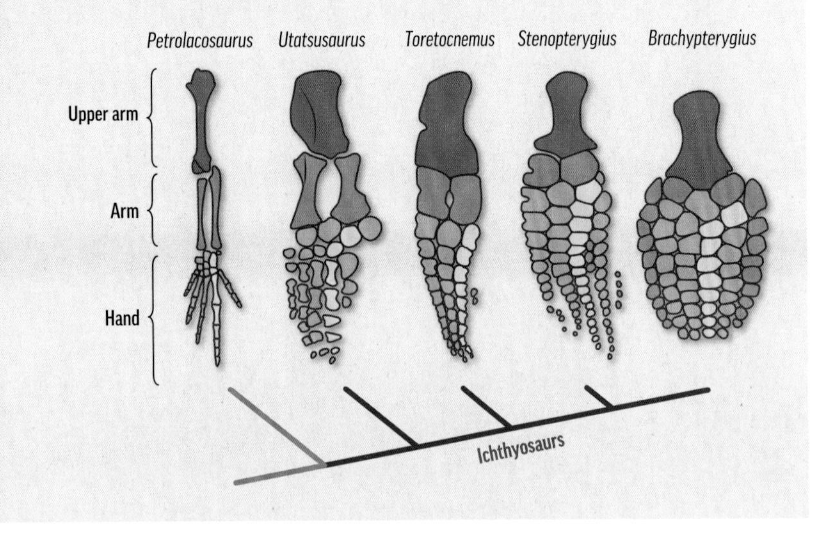

▲ Fig. 2.11. Different types of swimming paddles among ichthyosaurs: *Utatsusaurus*, Lower Triassic; *Toretocnemus*, Upper Triassic; *Stenopterygius*, Lower Jurassic; *Brachypterygius*, Upper Jurassic. *Petrolacosaurus*, a reptile from the Carboniferous, was chosen as the terrestrial representative of the group.

► Fig. 2.12. *Gulosaurus*, an ichthyosaur from the Lower Triassic in British Columbia (Canada).

▼ Fig. 2.13. (p. 36). *Eurhinosaurus*, an ichthyosaur from the Lower Jurassic in Europe.

phalanges, or bones in each finger), many later forms are characterized by significant hyperdactyly (a greater number of fingers) and hyperphalangy (see fig. 2.11 and "From Legs to Swimming Paddles," p. 39). *Platypterygius*, for example, had up to eight fingers and up to twenty rows of phalanges, all completely interconnected, forming a large and powerful, very rigid, paddle. Another example, *Ichthyosaurus*, had six fingers, each with more than thirty phalanges, the smallest of which resembled small, rounded tokens. Other ichthyosaurs were characterized by a *reduction* in the number of fingers: *Macgowania* and *Toretocnemus* had four and three fingers, respectively (nevertheless made up of a large number of phalanges); *Shastasaurus* (fig 4.11, pp. 126–27), the most "extreme" case, had a paddle formed by only two fingers. Finally, *Ophthalmosaurus* and *Brachypterygius* (fig. 2.11) had rounded paddles almost as wide as they were long, with six fingers but not many phalanges. So, the shape and structure of paddles varied considerably from one species of ichthyosaur to the next: they could be large, rounded or elongated, or even very long and slender. Judging by the extent of the traces of skin that have been preserved in some specimens, however, the swimming paddles were larger than the bones can show us. Paddles must therefore have been reinforced with **connective tissue**.

Only the most primitive ichthyosaurs had a pelvis that was still connected to the spine, as well as **stylopods** (the upper part of each limb) and **zeugopods** (the lower part of each limb) that were very long (fig. 2.11). In most ichthyosaurs, the pelvis and hind limbs had shrunk, although in some rare species the latter were almost the same size as the forelimbs. The size reduction of the rear paddles as compared with the front ones is correlated with a reduction in the number of fingers and phalanges; hyperphalangy and hyperdactyly occurred mostly in the front paddles.

In the *Guinness Book of Records*

It is among the ichthyosaurs that we find the largest eyes of any vertebrates, sometimes exceeding 25 centimeters in diameter! For comparison, the largest-ever vertebrate (the blue whale) has eyes that are a "mere" 11 centimeters in diameter. In the entire animal kingdom, only two kinds of cephalopods—giant squids (*Architeutis*) and colossal squids (*Mesonychoteutis*)—have eyes that are proportionally larger compared to the size of their bodies (fig. 2.14). The prize for the largest eyes relative to body size goes to (the aptly named) *Ophthalmosaurus* as well as to *Temnodontosaurus* (fig. 5.8, pp. 146–47). As in the case of the cephalopods of the ocean depths, these outsized eyes must have helped ichthyosaurs see in the dark when they dove far down.

You may be wondering how, given that the eyes are soft-tissue organs and do not fossilize, we can determine their size in ancient marine reptiles. Estimates of ichthyosaurs' eye size are made possible by the **sclerotic rings** that are preserved in the creatures' orbits (eye sockets). The rings, one in each orbit, are formed by little bones that overlap one another in a

Gulosaurus

This animal's name derives from a reference to the mining industry and to the economic development it gave rise to in the region of Lake Wapiti in Canada, where the 6-kilometer-long Wolverine tunnel allows Canada's West Coast to be supplied with coal. *Gulo* is the generic name for the glutton or wolverine in Canada: the name *Gulosaurus* therefore refers to the "wolverine lizard." This ichthyosaur, from the Lower Triassic (ca. 250 Ma) in British Columbia, is one of the oldest in its group. Phylogenetically it is very close to *Grippia*—and, like *Grippia*, its body must have been long and slender, with no dorsal fin. Unfortunately, in the case of *Gulosaurus* no specimens have been found in which any trace of the tail is preserved. Nonetheless, the specimens we do have allow us to observe that in this very ancient **taxon** the lower bones of the front limbs (i.e., the radius and ulna), which shrank as the ichthyosaurs evolved further, were long.

Eurhinosaurus

Eurhinosaurus, the "well-nosed lizard," is known from the Lower Jurassic in Europe (Germany, England, Belgium, France, and Switzerland). With an overall length of about 6 meters, it had a most impressive snout. The upper jaw projected well beyond the lower jaw, and it was equipped with teeth that pointed outward, giving them the appearance of swords. *Eurhinosaurus* had large swimming paddles and is considered to have been a very active swimmer capable of great bursts of speed, which may have helped it catch the fish that it ate. Whereas some paleontologists suggest *Eurhinosaurus* used its fearsome-looking muzzle to perforate or harpoon its prey, or to hurt its adversaries, most believe the prodigious snout was used to stir up sediment on the seabed while *Eurhinosaurus* looked for food there.

manner reminiscent of a camera diaphragm or iris (fig. 2.14; also see "Diving and Underwater Vision," pp. 150–51). They must have served a structural support function for those enormous eyes, as well as providing protection and resistance to pressure during deep dives. Sclerotic rings are present in many marine reptiles. Perhaps surprisingly, they can be found in numerous terrestrial animals, notably including both non-avian dinosaurs and birds (see "Diving and Underwater Vision," pp. 150–51).

As far as the aforementioned hyperdactyly and hyperphalangy are concerned, ichthyosaurs are the record holders in this area too, with up to ten fingers in each paddle and up to thirty phalanges for each finger. And ichthyosaurs came in an impressive range of sizes. While the smallest among them, such as *Chaohusaurus* (fig. 5.6, p. 143), were less than 1 meter long, the largest, such as *Shonisaurus* and *Shastasaurus* (fig. 4.11, pp. 126–27), could reach more than 20 meters.

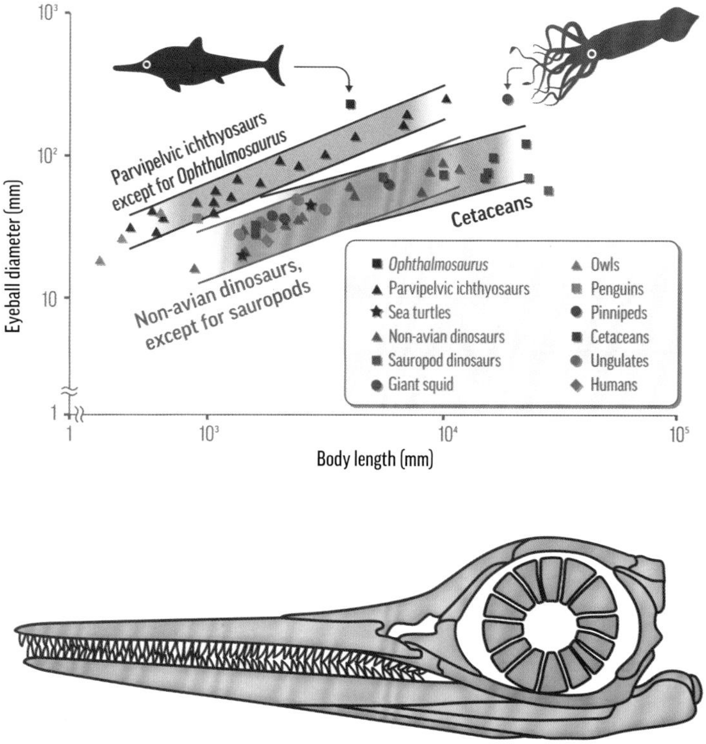

▲ Fig. 2.14. **Above:** comparisons of eye size in relation to body length in different types of animals. The ichthyosaur *Ophthalmosaurus* and the giant squid are on the steps of the podium.
Below: the skull of an ichthyosaur, showing the very large orbit and the sclerotic ring.

The Dress Doesn't Make the Dolphin

In illustrations, ichthyosaurs are often represented as having a dorsal fin and with the upper lobe of their caudal fin nearly as large as the lower lobe (a **hypocercal** tail), a profile that does somewhat resemble that of a dolphin or a shark. Of course, there is no relationship among ichthyosaurs, dolphins, and sharks, which belong to three very different classes of vertebrates (Reptilia, Mammalia, and Chondrichthyes, respectively). In addition, ichthyosaurs possessed a greatly varied set of physical characteristics, and each **taxon** exhibits adaptations to a specific ecology.

The oldest ichthyosaurs, such as *Utatsusaurus*, *Chaohusaurus*, *Gulosaurus* (fig. 2.12), and *Grippia*, have been found in Lower Triassic deposits in Spitsbergen (a Norwegian island), China, Japan, Canada, and Nevada (in the western United States). From an anatomical point of view, these primitive forms are clearly distinguishable from later ichthyosaurs (fig. 2.15). Their physique was long and slender; they lacked a dorsal fin but had a long, straight (or only slightly ventrally curved) tail, which suggests they swam by undulating their entire body, as eels do—in other words, using anguilliform locomotion (see "Locomotion in Ichthyosaurs and Sauropterygians," pp. 148–49).

The forms from the Middle and Upper Triassic—for instance, *Cymbospondylus*, *Mixosaurus* (fig 4.4), *Shonisaurus*, and *Californosaurus*—were anatomically much more varied, and geographically more widely distributed, than those of the Early Triassic. So, for example, while *Cymbospondylus* had a very elongated tail, later ichthyosaurs had a tail that had been transformed into a veritable caudal fin (with longer **neural** and **haernal spines**), as in the case of *Mixosaurus*, or a tail that was actually bilobed, as in the case of *Shonisaurus* (fig. 2.15). By the Middle Triassic, *Cymbospondylus*

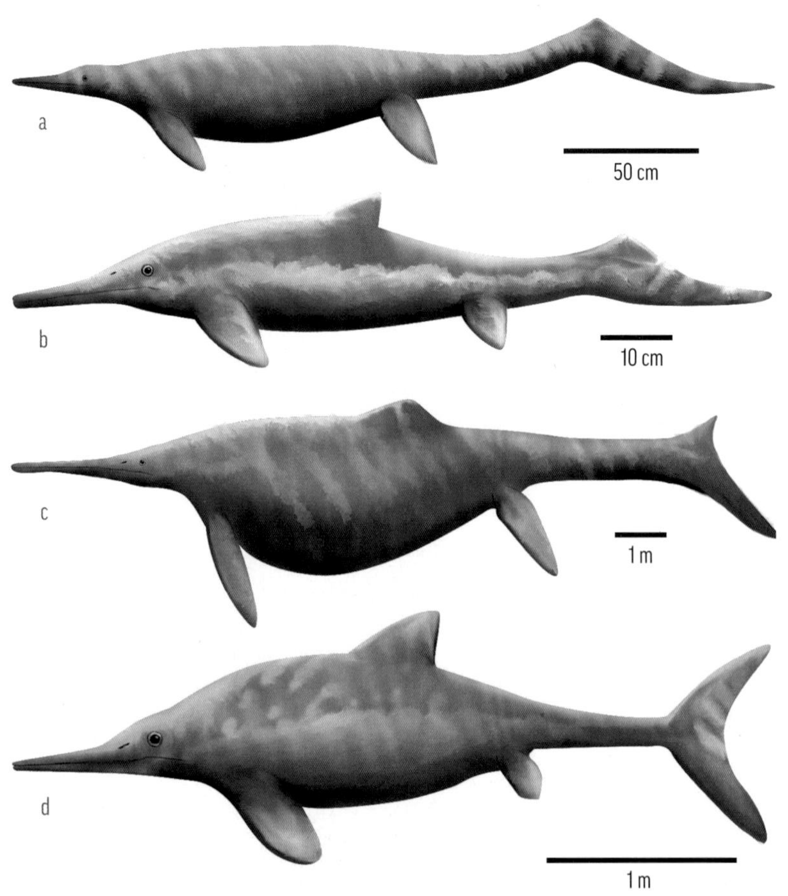

▲ Fig. 2.15. *Utatsusaurus* (a), *Mixosaurus* (b), *Shonisaurus* (c), and *Ophthalmosaurus* (d) exemplify differences in size and morphology among ichthyosaurs, particularly in the tail and the swimming paddles.

and other ichthyosaurs had begun to reach impressive sizes (the former, for example, being 10 meters long), and we have already mentioned the largest ichthyosaurs of the Late Triassic, *Shonisaurus* and *Shastasaurus.*

As far as the post-Triassic ichthyosaurs are concerned, for the most part they present similar physical characteristics (fig. 2.15): a rigid, tapered body; relatively small swimming paddles; and a bilobed (crescent-shaped) tail with a narrow connection to the rest of the body, reminiscent of some modern fish, such as lamnid sharks (the family to which the great white shark belongs), tuna, and mackerel, which are all rapid swimmers. (Unlike in those fish, in ichthyosaurs the spine was strongly curved ventrally and therefore the final bones of the tail supported the lower lobe of the caudal fin but not the upper lobe. However, this difference is no more than a detail and doesn't affect the properties of the caudal fin in all these forms in any way.) Typical of rapid swimmers that have developed a thunniform mode of swimming (this is a reference to tuna: see "Locomotion in Ichthyosaurs and Sauropterygians," pp. 148–49), such a construction permits a creature to propel itself using only its tail.

Many forms fell between the two "extremes" of aquatic locomotion (i.e., anguilliform and thunniform swimmers), with a gradual increase in the importance of the tail in propulsion and the progressive acquisition of true swimming paddles, shrinking of the pelvic bones, and development of a downward curvature in the final bones of the tail (and the growth of a fleshy upper lobe at the end of the tail).

Variation on the "Teeth of the Sea"

Our knowledge of ichthyosaurs' eating habits and modes of predation relies partly on the shape of their teeth and on their fossilized stomach contents.

Fossil ichthyosaurs exhibit a wide range of dental characteristics, which point to varied eating habits. Certain forms from the Lower and Middle Triassic are characterized by a **heterodont** dentition: in the back of the jaw they had teeth for crushing, and in the front they had conical teeth for perforating prey, as did *Grippia.* These generalist predators must have fed principally on shelled mollusks, notably cephalopods, but also on fish. Other ichthyosaurs, such as *Phalarodon,* were **durophages** (specialized to eat hard foods, such as heavily armored creatures). Their short, bulbous teeth must have allowed them to crush tough prey efficiently. And finally, some Triassic forms, such as *Shastasaurus* (fig. 4.11, pp. 126–27) and *Shonisaurus,* must have sucked up their prey much as modern beaked whales do—because, like beaked whales, they were either toothless or nearly so. Post-Triassic ichthyosaurs had teeth that were conical and pointed, allowing for a diet of fish or meat. These animals—generally mega-predators like

From Legs to Swimming Paddles

The secondary adaptation to an aquatic environment was always accompanied by significant changes to the limbs (fig. 2.16). Among amphibious forms, such as crocodiles, such changes were relatively limited, because the animals still needed to be able to move about on land. Among forms that were exclusively aquatic but were not very active swimmers, which moved through the water mostly with the aid of their four limbs (this includes nothosaurs, placodonts, and certain crocodyliforms), the hands and feet were generally palmed, but the proportions of the different parts of the limb were either not at all or only slightly changed. On the other hand, the swimming paddles of the most active swimmers (ichthyosaurs, mosasaurs, plesiosaurs, sea turtles, and cetaceans) are usually characterized by changes in proportions: a reduction in the relative size of the bones of the upper part of the limbs (from upper arm/thigh to forearm/lower leg), accompanied by a reduction in the number of bones in the wrists and ankles and an increase in size as well as a lengthening of the extremities (hands/feet). Such paddles are involved in locomotion to a varying degree.

Although plesiosaurs and ancient sea turtles depended on their paddles for propulsion, mosasaurs and ichthyosaurs, which had particularly small hind limbs, used their paddles mostly for balance and maneuverability. An excess of fingers (i.e., more than five), also known as hyperdactyly, allowed certain groups of marine reptiles to have paddles that were larger. An increase in the number of phalanges (i.e., the number of bones in each finger), or hyperphalangy, which facilitated an elongation of the paddle, was minimal in the case of most mosasaurs but present to a remarkable degree in certain ichthyosaurs and plesiosaurs.

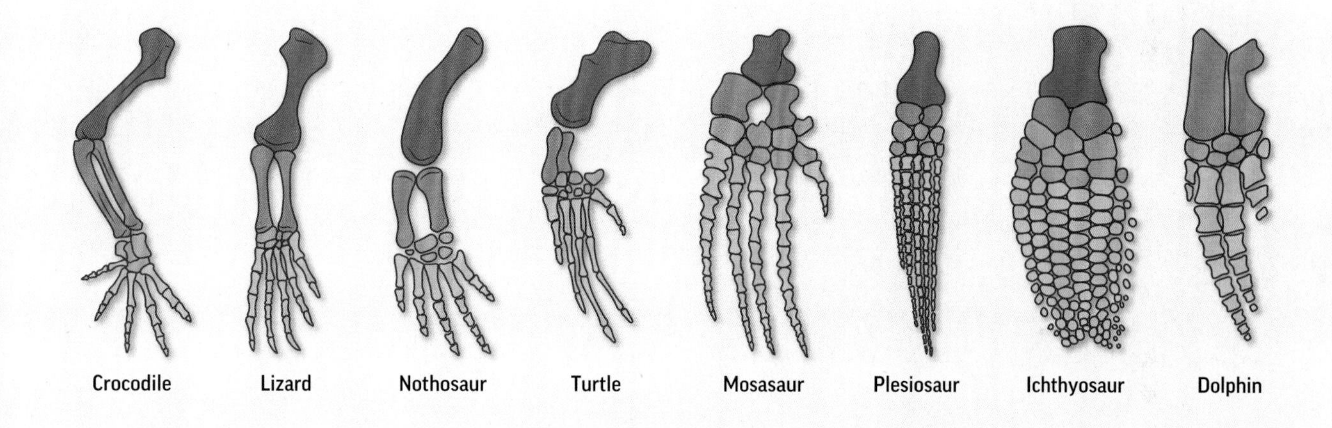

▲ Fig. 2.16. Different kinds of swimming paddles in marine tetrapods (reptiles and mammals) compared to the limbs of terrestrial reptiles (crocodiles and lizards).

Temnodontosaurus (fig. 5.8, pp. 146–47)—must have fed on other marine reptiles.

These differences in feeding habits are also indicated at the level of the anatomy of the skull. Durophagous ichthyosaurs had a shorter, smaller, more robust, and slenderer muzzle than forms that fed on fish, while megapredators' muzzle was generally long but robust.

Because of their overall shape and their swimming techniques, which we presume were anguilliform (see "Locomotion in Ichthyosaurs and Sauropterygians," pp. 148–49), the generalist and durophagous ichthyosaurs are presumed to have been ambush predators, hunting from lookout positions in shallow waters. They disappeared before or at the end of the Middle Triassic. On the other hand, ichthyosaurs with a thunniform swimming technique from the Upper Triassic, and all the forms from the Jurassic and Cretaceous, were more pelagic and were probably pursuit predators.

Mary Anning, One of the First Palaeontologists

◄ Fig. 2.17. Mary Anning and her dog Tray, at some point prior to 1842, to the east of Lyme Regis.

From the end of the seventeenth century on, several fossilized vertebrae, which we now know belonged to ichthyosaurs and plesiosaurs, started attracting the attention of naturalists but were thought to belong to "fish," a term which then included a number of marine animals, including cetaceans. While for a long time naturalists, thinkers, and even artists (like Leonardo da Vinci) had been asking themselves about various such petrified objects (called "osteoliths"), which had been found since antiquity, no serious hypothesis had been formulated to explain their true nature, their often truly ancient age, and the reason for their presence in rocks. According to the religious precepts that were prevalent at the time, Earth was estimated to be on the order of six thousand years old, so fossils were most often interpreted as being the remains of animals that had drowned in the biblical Flood. However, in the first decades of the nineteenth century, there arose two new scientific disciplines: geology (thanks to the Scotsmen James Hutton and Charles Lyell) and paleontology (thanks to the Frenchmen Jean Baptiste de Lamarck and, above all, Georges Cuvier—see p. 95). Developments in these fields would lead to considerable advances in our knowledge and interpretation of fossils.

Like Maastricht in the Netherlands (see p. 95), Lyme Regis, a small seaside resort in Dorset, on the southern coast of England, is a major destination for paleontologists, especially those specializing in marine reptiles of the Mesozoic. This small town is also the location John Fowles chose for his novel about the love between Sarah Woodruff and paleontologist Charles Smithson in *The French Lieutenant's Woman*, made famous following its 1981 movie adaptation of the same title, directed by Karel Reisz. Lyme Regis is also the center of the action in Tracy Chevalier's novel *Remarkable Creatures* (2009), adapted for the screen by Francis Lee as *Ammonite* (2020). Why this location for these love stories? What is the common bond that ties them together? The answer is fossils—numerous, spectacular fossils, of historical interest.

Lyme Regis's fossils would not mean much if behind them there were not a woman, Mary Anning (1799–1847), who remained in the shadows for over a century and a half but has recently found her place in the limelight. She lived to be only forty-eight, in an England in which little consideration was given to women who were born poor. She nevertheless left an indelible mark on the development and the influence of paleontology, especially on the exciting history of the discovery of marine reptiles from the Mesozoic.

The history of this woman is a singular one, starting with a twist of fate when, at the age of 15 months, she was the only survivor in a group of people hit by lightning. Her father was a carpenter, and the family had difficulty making ends meet. To round out their income, he explored the imposing coastal cliffs of black marl that border Lyme Regis to the east and the west (fig. 2.18), looking for fossils. From a very early age, Mary and her brother Joseph would accompany their father, while their mother sold their findings to tourists. Fossils of invertebrates are abundant in the thick marly series of the local Blue Lias formation, deposited in the early Early Jurassic (roughly 200–190 Ma), but the remains of vertebrates (such as vertebrae and teeth) are found here as well.

Around 1810 Joseph found a complete skull, 1.2 meters long, of an animal that was unknown at the time. Several months later, thanks to a landslide at the same location, Mary (who was 12 years old) and her brother found the torso corresponding to the skull. The animal must have been almost 5 meters long. It was the first mostly complete skeleton of an ichthyosaur ever found. But for the time being, no one was aware of it, and for a good reason: this emblematic group of marine reptiles from the Mesozoic would not be identified as such until several years later. For the moment, scientists attributed the skeleton to an unknown "fish."

This discovery was the beginning of a long career for Mary, who would establish her reputation as a "fossil searcher" throughout Europe. The many skeletons she uncovered allowed for the description of the first ichthyosaurs and plesiosaurs, by English geologists William Conybeare and Henry de la Beche, between 1821 and 1824. These authors described at least three species of ichthyosaurs from Lyme Regis: *Ichthyosaurus communis* (3 meters), *I. tenuirostris* (today *Leptonectes*, 4 meters) and *I. platydon* (today *Temnodontosaurus*, 9 meters), a mega-predator from the Lower Jurassic (see chapter 5, pp. 146–47). The first specimen Mary and her brother had found was an *I. platydon.*

When in 1823 Mary was the first to discover a plesiosaur, *Plesiosaurus dolichodeirus* (see "*Plesiosaurus*," p. 55), the creature's form seemed so improbable that even important scientists such as Cuvier surmised the specimen was a fake, assembled by Mary from fossils of different origins. But after other skeletons exhibiting the same characteristics were found, Cuvier had to surrender to the evidence and admit his mistake. Mary also discovered several thalattosuchian crocodiles; was the first to find a pterosaur outside Germany; discovered several remarkable actinopterygians; found fossils of several previously unknown cephalopods, some with their ink sac preserved; and was the first to find fossilized excrements, called coprolites.

But Mary Anning was not only a fossil "hunter." Although she had received only a minimal education, she nonetheless acquired, thanks to an insatiable curiosity and by consulting the works available at the time, a deep knowledge of the fossils she unearthed and the terrain she found them in. She also knew how to draw and interpret them. The extent of her knowledge

▼ Fig. 2.18. The Blue Lias cliffs, east of Lyme Regis, county Dorset (England), where Mary Anning found her Lower Jurassic fossils. These cliffs are part of the Jurassic Coast, Britain's only World Heritage Site.

◄ Fig. 2.19. Henry de la Beche, *Duria Antiquior*, 1830. National Museum Cardiff.
This is the first reconstruction of a complete paleoecosystem, based on the discoveries made by Mary Anning in the Lias of Lyme Regis.

rivaled that of established researchers, and her understanding of the terrain had few equals: many learned men came to consult her and to look for fossils with her. Swiss American paleontologist Louis Agassiz, impressed by his meeting with Mary, would name two species of fossil fish in her honor (*Acrodus anningiae* and *Belenostomus anningiae*).

Although her exceptional discoveries were the basis for the careers of many scientists of the period, she was never quoted in their articles, and almost no one, with the exception of Louis Agassiz, made reference to her discoveries. She never published an article under her name alone, except for a letter she sent to the *Magazine of Natural History* in 1839 to point out that she had discovered the newly named fossil shark *Hybodus* long before its naming, by Louis Agassiz, in 1837. For her, *Hybodus* was therefore not "new," as the magazine described it. The letter did perhaps betray some naivete on her part, in that in paleontology (and biology) a species does not officially exist until after it has been named and described.

As a woman born into the working class, Mary had only very restricted educational opportunities, and access to the scientific circles of the period, such as the Geological Society of London, was forbidden to her. Nevertheless, the academic world showed its gratitude because, thanks to the kindness of her friend, paleontologist William Buckland, in 1820 the British Society for the Advancement of Science granted her an annual pension. Likewise, when in 1847 she fell gravely ill, the Geological Society of London organized a subscription to help pay her expenses. She remained poor nonetheless, even after the sales she made of her many finds. The time when important fossils would sell for millions of dollars at auction houses had, unfortunately for her, not yet arrived. Many of her longtime friends came to her financial aid several times during her life. In 1820, Colonel Thomas Birch sold fossils that he had bought previously from her at an auction organized in London and donated the proceeds to her; Georges Cuvier was present at this sale and acquired several specimens of ichthyosaurs and plesiosaurs for the museum in Paris. In 1830 Henry De la Beche completed the first reconstruction of the history of an ecosystem (named *Duria Antiquior* [fig. 2.19], which means "A More Ancient Dorset"), based on the fossils found by Mary Anning; the profit from the sales of the copies he had made of it were intended to help her. After her death, the members of the Geological Society paid tribute to her and had some stained glass created in her memory placed in the church of Saint Michael in Lyme Regis, where it can be seen today. Nevertheless, as time went by her name was slowly forgotten, but in the last several decades she has been remembered and rightly given consideration as an important figure in paleontology. In 2010 members of the Royal Society named her among the ten women in British history who have had the greatest influence on science.

Mary Anning's exceptional discoveries had a major impact on the development of paleontology as a science and stimulated the learned men of the time in all respects. The often complete and well-preserved skeletons she found allowed researchers, from the 1820s on, to form a precise idea of what ichthyosaurs and plesiosaurs must have looked like, to understand their reptilian nature, and to realize they had no equivalent in the nature of the period. Thanks to these new fossils, Cuvier's theory—according to which, worlds inhabited by creatures very different from today's must have existed and then disappeared—became increasingly convincing.

The Sauropterygians: Loch Ness Monster & Co.

The name "sauropterygian" (from the Greek *sauros*, "lizard," and *pterux*, "wing," a reference to the anatomy of their swimming paddles) is hardly known to the public, and yet the order Sauropterygia includes one of the Mesozoic's most recognizable groups of marine reptiles: the plesiosaurs. These animals with a long neck and a barrel-shaped body, which some optimists imagine they have glimpsed in the troubled waters of Loch Ness, fascinate people, because their bizarre construction has no modern or past equivalent. Long before the plesiosaurs, however, other sauropterygians swarmed in the seas and oceans of the Triassic, not suspecting that they would someday lose their place to these monstrous-looking relatives.

It was British paleontologist Sir Richard Owen who, in 1860, proposed the name Sauropterygia to describe a group of animals that included, in addition to the plesiosaurs, the nothosaurs and the placodonts. Although our knowledge has grown considerably since then, the label "sauropterygian," after having fallen out of use for a very long time, started to regain favor at the end of the 1990s and is today used by everybody in the field, which proves that Owen's observations were sufficiently accurate for his definition of the group to remain valid.

And yet the kinship relations between Sauropterygia and other groups of reptiles are not yet clearly established. Sauropterygians are most often considered diapsids that lost

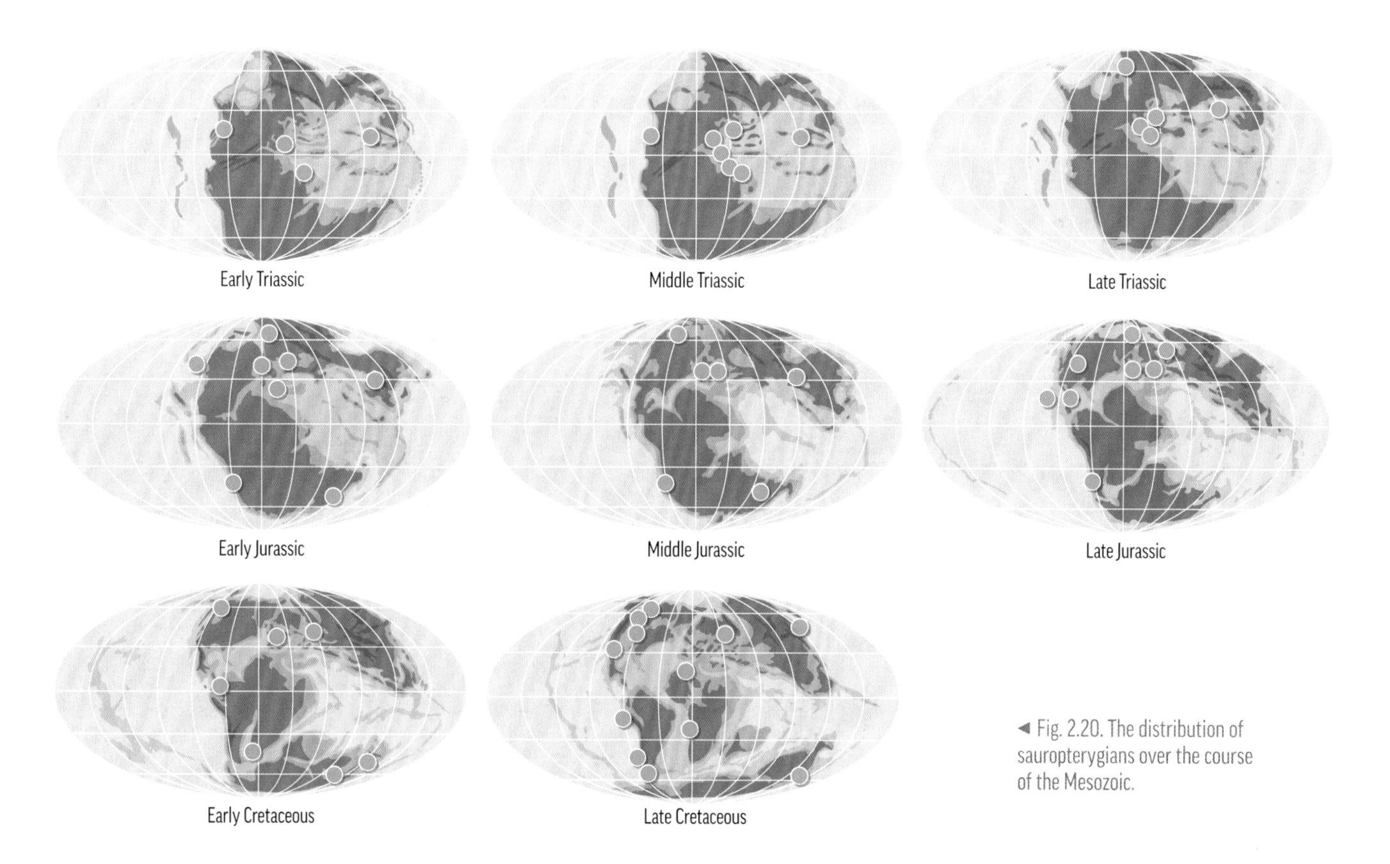

◄ Fig. 2.20. The distribution of sauropterygians over the course of the Mesozoic.

their lower temporal fossae. After appearing at the beginning of the Triassic, they divided into two major lineages: the eosauropterygians, which include the famous plesiosaurs and pliosaurs, and the placodonts, which had very atypical characteristics. The most diverse of all marine reptiles, sauropterygians had spread throughout the world by 180 million years ago (fig. 2.20).

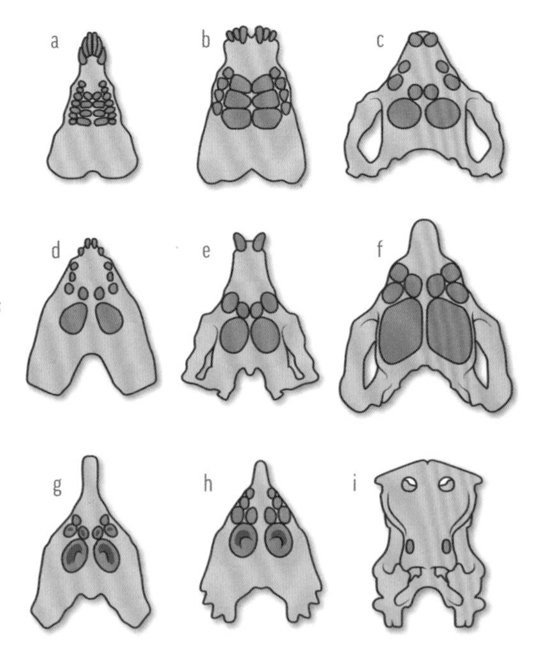

► Fig. 2.21. A palatal view of the skulls of various placodonts, showing the large, short, bulbous teeth that are typical of the group as well as their differences in form and arrangement. *Paraplacodus* (a), *Placodus* (b), *Cyamodus* (c), *Sinocyamodus* (d), *Protenodontosaurus* (e), *Macroplacus* (f), *Psephoderma* (g), *Placochelys* (h), *Henodus* (i).

The Placodonts: Have the Armor-Clad Step Forward!

During the Triassic, over the course of about 50 million years (252–201 Ma), the placodonts lived in the epicontinental seas at the edges of the Tethys (see chapter 1, p. 10). Their fossilized remains have been found

Henodus

Henodus, a placodont, comes from the Lower Carnian stage in Germany. Roughly 1 meter long, it was almost equally wide. Its back and its underside were armored by bony plates of various sizes—called osteoderms—and covered with horny scales. Its rectangular, flattened head was significantly different from that of most placodonts. The sharp borders of its snout were lined with tiny denticles. In its actual jaws were only four small teeth, two on the top and two on the bottom (fig. 2.21i). The shape of its jaws, and its notable lack of teeth, suggest that the power of its bite and its crushing capability were considerably less than those of other placodonts. *Henodus* may in fact have been a suspensivore (filter feeder).

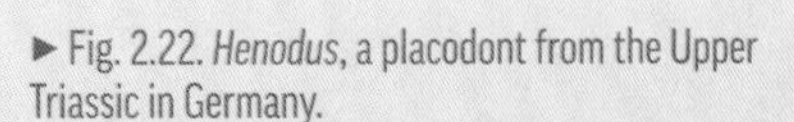

► Fig. 2.22. *Henodus*, a placodont from the Upper Triassic in Germany.

mainly in Europe and China. The name "placodont," which means "flat tooth," derives from the fact that most of these animals were equipped with large teeth arranged like cobblestones on the palate and the lower jaw, which suggest durophage eating habits (fig. 2.21).

To supplement this cobblestone dentition, some placodonts, such as *Placodus* and *Palatodonta* (from the Netherlands), were endowed with front teeth that were cone-shaped or with incisors, which they probably used to grab their food. Traditionally it is thought that placodonts fed on hard-shelled mollusks and crustaceans. So *Placodus*'s massive skull (fig. 2.23), which in certain places exhibits signs of pachyostosis (see "The Secrets of Bone," p. 129), and its robust mandible must have been meant to efficiently resist the strong pressures experienced while crushing tough prey. Certain placodonts, however, such as *Henodus* (fig. 2.22), had very few teeth, which leads paleontologists to suppose they sucked up their prey instead of catching it in their teeth.

Placodonts are divided into two groups, according to whether they exhibit body armor or not. The cyamodontoids do. The placodontoids don't or were endowed only with minimally developed bony structures in a line along the spine (fig. 2.24, top). *Placodus*, for example, displays a unique row of bony plates and belongs to the second group. The bony plates of the cyamodontoids were sometimes so developed and contiguous that they covered the entirety of their body and made them resemble turtles, as is the case with *Henodus* and *Plachochelys* (*chelys*

Placodus

Placodus is one of the two genera of the placodontoids. On the upper jaw of its massive skull, it had six front teeth shaped like incisors; it had eight bulbous teeth on the sides; and it had six large and flattened teeth, like cobblestones, on the roof of its mouth (fig. 2.21b). It would dig up the seabed looking for invertebrates, which it grabbed with its front teeth before crushing them with its rear ones. Almost 2 meters long, *Placodus*'s body was not covered by a shell; it was outfitted only with a median range of bony plates that overhung the spine (fig. 2.24).

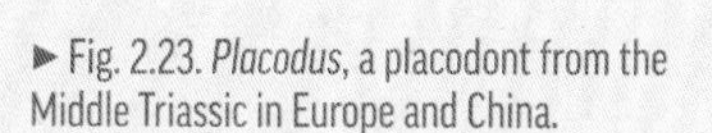

► Fig. 2.23. *Placodus*, a placodont from the Middle Triassic in Europe and China.

▲ Fig. 2.24. Different types of body armor exhibited by placodonts (*Placodus*, *Cyamodus*, *Psephoderma*), from a simple dorsal row of osteoderms among placodontoids to a strong body armor with diverse and varied ornamentations among the cyamodontoids.

means "turtle" in Greek). In fact, cyamodontoids were once thought to be close relatives of turtles before being considered more properly as sauropterygians.

Cyamodontoids and turtles are the only amniotes ever to have developed armor covering the back, the belly, and the sides. Nevertheless, while all cyamodontoids had a dorsal shield (on the back), the ventral one (on the belly) could be underdeveloped or even absent. The dorsal shield could consist of a single support, as in *Henodus* and *Placochelys*, or be made up of two parts, the larger one situated frontally while the smaller one covered the pelvic area, as in *Cyamodus* (fig. 4.4, pp. 116–17) and *Psephoderma* (fig. 2.24). In addition, while both cyamodontoids and turtles were or are armored, their armatures do not have the same origin at all.

Turtles' shell, made up of a dorsal section called a carapace and a ventral section called a plastron, develops mostly starting from the bony elements of the skeleton: neural spines (bony dorsal projections of the vertebrae) and ribs that are more or less welded together for the carapace; clavicle and ventral ribs for the plastron (see fig. 2.56). Cyamodontoids' armature, by comparison, was generally an assemblage of extra bones that developed in the skin (osteoderms) and formed numerous small plates, either flat or pointed in shape, with a round, polygonal, or hexagonal base. Their armature therefore did not contain any skeletal elements. Because of their dermal origin, these plates were more closely related to those of certain archosaurs, such as the crocodylomorphs (see p. 66) and some dinosaurs (e.g., ankylosaurs and titanosaurs). When cyamodontoids were alive, they must have been covered in keratin. This is the only point they share with turtles, since the plates of turtles' shells are also covered in horny, keratinized scales.

Placodonts' unique set of features has led to many questions regarding their ecology. The placodonts are considered the "sea cows" of the Triassic, or as having engaged in behaviors resembling those of the sirenians (today's dugongs and manatees; see chapter 7, p. 196). And the skeletons of sirenians and those of numerous placodonts do show anatomical and functional **convergences**: they exhibit a marked bone density, called **pachyosteosclerosis** (see "The Secrets of Bone," p. 129), one consequence of which is to make the animals heavier, thus allowing

▲ Fig. 2.25. Placodonts supposedly led a life of digging for food on the seabed.

them to remain submerged near shallow seabeds with minimal muscular exertion and thus minimal use of energy. Among sirenians, this increase in bone mass can be seen especially in the rib area.

In the case of numerous placodonts, the shell or the gastralia are what mostly weighed them down, but this increase in the mass of certain bones did sometimes increase the ballasting effect. Placodonts were therefore probably able to station themselves on the seafloor while searching for food, thanks to their ballasted skeleton (fig. 2.25). Being encumbered with such massive and heavy bodies, they very likely were not fast swimmers.

The Eosauropterygians: Great Travelers

The group Eosauropterygia comprises all sauropterygians other than the placodonts. Eosauropterygians first appeared during the Early Triassic and did not go extinct until the Cretaceous/Paleogene crisis, thus demonstrating an unparalleled longevity among marine reptiles. They occupied all the world's seas and oceans.

Eosauropterygians comprised several lineages: the pachypleurosaurs, the nothosaurs, and the pistosaurs (among which we find the famous plesiosaurs and pliosaurs). Paleontologists often speak of "Triassic sauropterygians" to refer to the sauropterygians other than plesiosaurs and pliosaurs—in other words, those groups that did not continue into the Jurassic. The evolutionary history of Triassic sauropterygians seems tied to fluctuations in sea level: the rise in sea level during the Triassic allowed for the establishment of shallow epicontinental seas, in which these reptiles multiplied and diversified.

The Pachypleurosaurs

The pachypleurosaurs (from the Greek *pachy*, "thick"; *pleuro*, "rib"; and *sauro*, "lizard") are known only from the Middle Triassic and the beginning of the Upper Triassic. Their fossilized remains have been found in western Tethysian provinces (Europe) and eastern ones (China). They were generally small, with the largest of their dozen species barely exceeding a meter in length, although

▲ Fig. 2.26. A cast of *Keichousaurus*, a pachypleurosaur from the Triassic in China. The largest specimens are only 30 centimeters long and must have measured about 5 centimeters at birth.

some, such as *Wumengosaurus*, reached nearly 1.5 meters.

Pachypleurosaurs all exhibit very similar characteristics: a long body, neck, and tail, with a small skull and a short snout. They had large eye sockets and relatively small temporal fossae, the opposite of nothosaurs. Their limbs still resembled those of the terrestrial animals they evolved from, with long zeugopods and five fingers that exhibit no hyperphalangy (compare "From Legs to Swimming Paddles," p. 39).

Most pachypleurosaurs—for instance, *Keichousaurus*—exhibit pachyosteosclerotic ribs (see "The Secrets of Bone," p. 129). We imagine that their ballasted skeleton allowed them, like the placodonts, to linger near the seabed for relatively long stretches without needing to spend much energy. Pachypleurosaurs were therefore small reptiles that lived in shallow waters and that probably moved fairly slowly.

Keichousaurus (fig. 2.26) is probably the most famous pachypleurosaur. Numerous adults and juveniles of this species, 5–25 centimeters long, were discovered in China starting in the 1950s. The specimens are generally complete and allow for detailed studies of anatomy, most notably variations within a single species. This is how paleontologists were able to observe that in some specimens, the humerus (upper front limb bone) and femur (upper hind limb bone) were the same size, while in others the humerus was both more massive and longer than the femur. It was hypothesized that this was a case of sexual dimorphism ... but which specimens were the males and which were the females? The answer came in 2004, when a Chinese team published an essay accompanied by photographs of fossils of adult *Keichousaurus* containing embryos. And these pregnant *Keichousaurus* had a humerus and a femur that were the same size. For a long time, it had been supposed that sauropterygians, which were probably completely independent of the terrestrial environment, gave birth to live young, but never had a pregnant sauropterygian been found, so proof of viviparity had been lacking. Importantly, each gravid female discovered by the Chinese team contained between four and six embryos at

Simosaurus

Simosaurus ("lizard with a blunt snout") was a Middle Triassic nothosaur 3 to 4 meters long, found mostly in Europe. Its large and relatively flat skull lacked the long muzzle of *Nothosaurus*, and its blunt teeth support the notion that it subsisted on a diet of tough foods. The genus *Simosaurus* comprises a single species, *S. gaillardoti*, first described by German paleontologist Herbert von Meyer in 1842.

► Fig. 2.27. *Simosaurus*, a nothosaur from the Middle Triassic in Europe.

the level of the rib cage, just ahead of the pelvic girdle. *Keichousaurus* must have therefore given birth, in the water, to young that were already formed and capable of swimming. This adaptation shows that sauropterygians became independent of a terrestrial environment very early in their evolutionary history.

Coastal Fishers: The Nothosaurs

The nothosaurs inhabited the Tethysian and Pacific seas of the Triassic. They weren't very diverse and are represented by only four genera. Their skull was longer and slenderer than pachypleurosaurs', and their temporal fossae (see p. 25) were larger than their eye sockets. Overall, nothosaurs were larger than pachypleurosaurs as well, even though one species (*Nothosaurus winkelhorsti*) had a skull only 5 centimeters long.

Nothosaurs were probably piscivorous but could also feed on soft-shelled invertebrates. Their slender jaws, with thin, pointed teeth, like those of *Lariosaurus* (fig. 2.28), probably did not allow them to latch on to tougher prey. There are nevertheless always exceptions: *Simosaurus* (fig. 2.27), with its shorter rostrum (another word for snout or muzzle) and more robust teeth, could certainly feed on invertebrates with harder shells.

▲ Fig. 2.28. *Lariosaurus*, a nothosaur from the Middle Triassic in the eastern Pyrenees (France), paleontology collections of the Sorbonne University (Paris, France). This skull, as fine as lace, is only 6 centimeters long.

Because, like pachypleurosaurs, nothosaurs had long bodies and limbs resembling those of terrestrial animals, they seem unlikely to have lived in the open ocean. In addition, in an open environment, where food resources might be spread out across long distances, good acceleration and maneuverability are indispensable. The high bone mass that is characteristic of some nothosaurs would have hampered them in these regards. It is therefore probable that most of these species were not very active swimmers and lived near the coast in shallow marine environments.

A Collection of Stars: The Pistosaurs

The pistosaurs, just like the nothosaurs, lived in the Tethysian and Pacific seas of the Triassic. Considered the closest relatives of the plesiosaurs and pliosaurs, they are known for six genera, among them *Yunguisaurus* (fig. 2.29) and *Bobosaurus*. The latter has, however, been recently placed among the plesiosaurs; its kinship relations are still debated.

The first pistosaurs do not seem to have varied much, in comparison with the plesiosaurs. They have, however, been found widely—principally in Europe, but also in China and the United States. They are known from the Lower Triassic to the Upper Triassic and were of respectable proportions, with *Bobosaurus* reaching about 3 meters in length.

From "Sea Giraffes," the Plesiosaurs, to the "Teeth of the Sea," the Pliosaurs

Phylogenetically, the plesiosaurs are part of the large clade of the pistosaurs (fig. 2.30). But unlike the Triassic pistosaurs, the plesiosaurs (from the Greek *plesios*, "close to," and *sauros*, "lizard") thrived for most of the Mesozoic. They are distinguishable from Triassic pistosaurs by several characteristics of their skeletons—for example, hyperphalangy and the shape of the radius and the ulna (the two bones in the lower part of each forelimb).

The earliest plesiosaur fossils date from the Upper Triassic. They were found in the United Kingdom and in Russia, in sediments roughly 225 to 215 million years old. However, it is believed these creatures differentiated themselves about 10 to 20 million years earlier, in the geologic age known as the Carnian, around 235 million years ago. After enduring for more than 150 million years, the plesiosaurs disappeared (as did the non-avian dinosaurs) during the famous Cretaceous/Paleogene crisis.

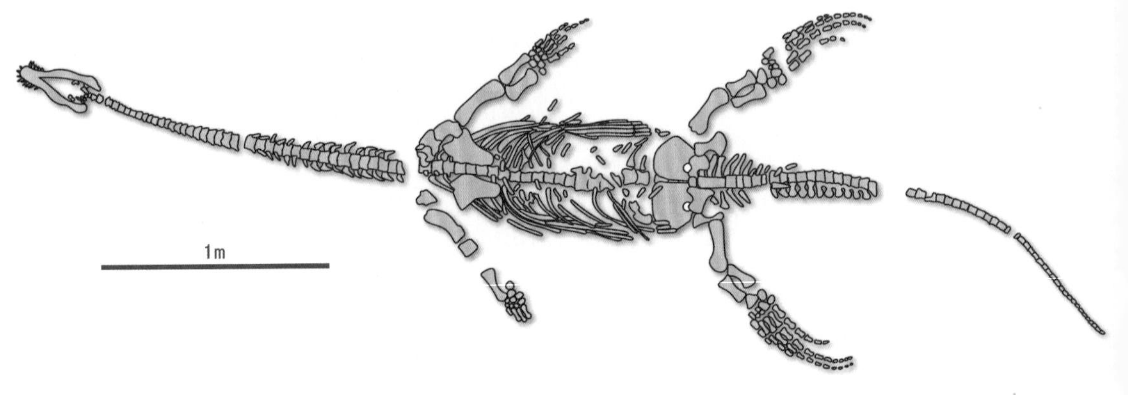

► Fig. 2.29. *Yunguisaurus*, a pistosaur from the Middle Triassic in China. Its anatomy indicates that it had evolved past needing to undulate its body in order to swim; it did, however, still have a long tail, like anguilliform swimmers.

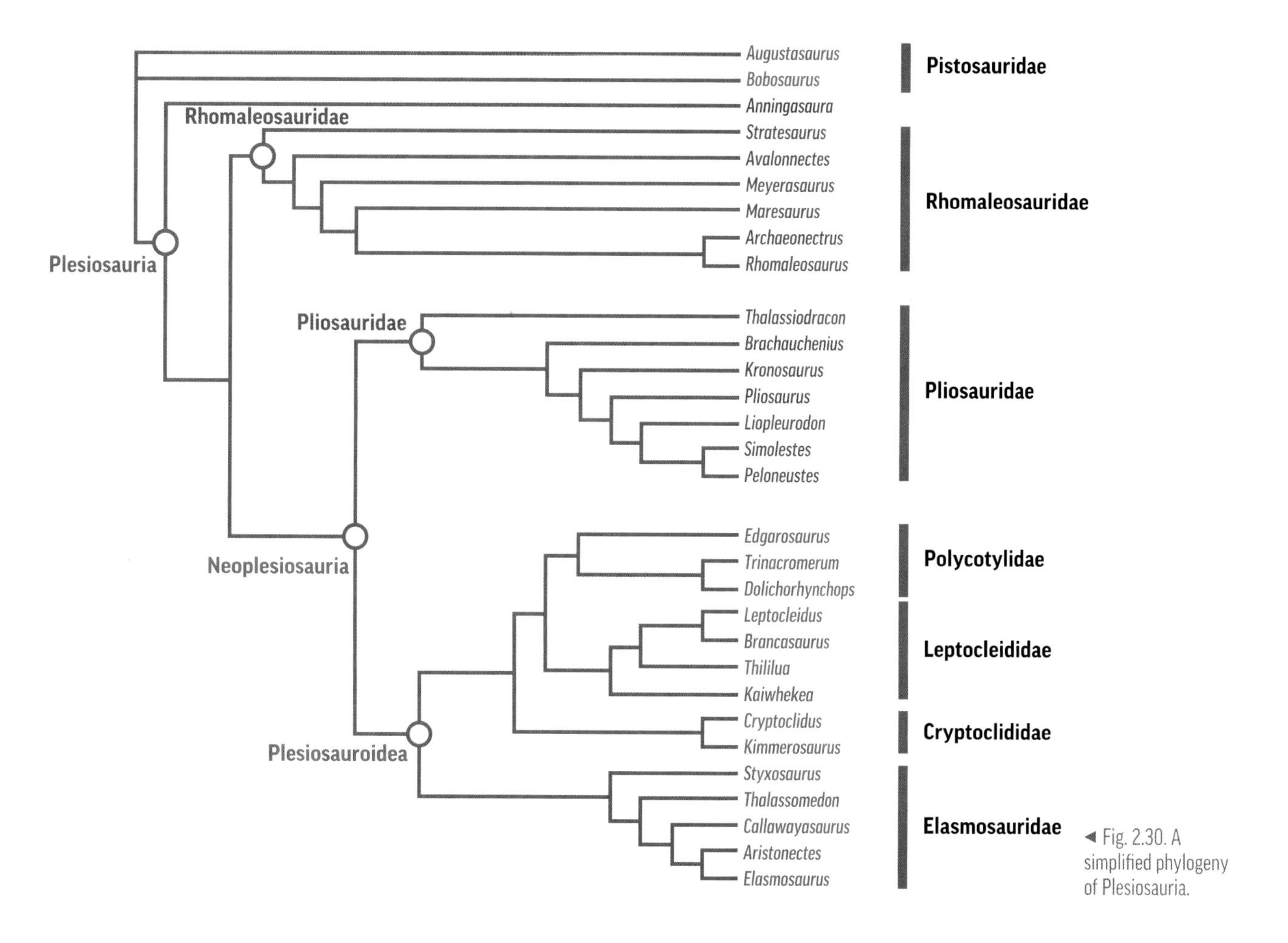

◄ Fig. 2.30. A simplified phylogeny of Plesiosauria.

The plesiosaurs (suborder Plesiosauria) (fig. 2.30) are traditionally, according to their physical features, divided into two groups: plesiosaurs and pliosaurs. Plesiosaurs thus in the narrowest sense of the term correspond to forms with a small head and a long neck; pliosaurs are those with a massive skull and a short neck. (But we shall see that looks can be deceiving.)

Plesiosauria was one of the most diverse groups of aquatic vertebrates. About one hundred species are known. Having spread worldwide starting in the Early Jurassic, members of Plesiosauria have been found fossilized on all continents and at all latitudes, from the Antarctic to the Artic. This suggests that they were active swimmers, capable of migrating over long distances. Their fossilized remains are abundant in marine deposits; in addition, fossils in England, Canada, and even Australia have been found in sediments of continental origin. These deposits correspond to ancient lagoons, rivers, or deltas; because the remains are mostly those of juveniles, some researchers have ventured that these shallow environments functioned as plesiosaur nurseries.

These highly successful marine reptiles were characterized by a massive body, four swimming paddles, and a long neck, although neck length was somewhat variable. In terms of overall length, they ranged from less than 2 meters for *Thalassiodracon* to 10 meters or more for the largest pliosaurs, such as *Kronosaurus* and *Liopleurodon* (roughly 10 meters), *Thalassomedon* (12 meters), and *Pliosaurus* (15 meters!). The smallest among them, such as *Thalassiodracon* and *Avalonnectes*,

▲▼ Fig. 2.31. **Above**, *Rhomaleosaurus*, from the Lower Jurassic in England. This gigantic rhomaleosaurid was more than 7 meters long. **Below**, *Meyerosaurus*, a rhomaleosaurid from the Lower Jurassic in Germany. Diagram of a skeleton preserved in the Staatliches Museum für Naturkunde (Stuttgart, Germany).

Green–pectoral girdle,
Pink–pelvic girdle,
Red–ribs,
Yellow–gastralia,
Blue–humeri,
Orange–femurs

abounded at the very beginning of the Jurassic but, for the remainder of the Jurassic and the Cretaceous, slowly gave way to animals of much larger size.

Unlike those of other sauropterygians, plesiosaurs' limbs were deeply modified and extremely well adapted to the work of swimming. These creatures propelled themselves not with spine or tail movements but by using their swimming paddles in underwater flight (see "Locomotion in Ichthyosaurs and Sauropterygians," pp. 148–49). The front and rear swimming paddles were fairly similar, although plesiosaurs in the narrow sense of the term generally had longer front swimming paddles, whereas pliosaurs generally had longer rear ones, which is a good way to distinguish between them. The upper bones of the limbs (the humerus and femur, respectively) were the longest bones, while zeugopod (radius and ulna; tibia and fibula) length was reduced; these disc-shaped elements were not elongated as in the Triassic sauropterygians. There were five fingers to a swimming paddle, and different species had different degrees of hyperphalangy. Unlike in ichthyosaurs, among plesiosaurs there was no hyperdactyly (see "From Legs to Swimming Paddles," p. 39). The surface area of the swimming paddles, which were not very flexible, increased considerably thanks to the increase in the number of phalanges, and the paddles' efficacy for locomotion was assured by strong muscles connecting the front paddles to the large pectoral girdle and connecting the rear ones to the pelvic girdle (fig. 2.31, bottom). As in ichthyosaurs, the

underside of the rib cage was reinforced by gastralia. Last, their tails were shorter in comparison with Triassic sauropterygians.

Some species possessed a long and slender snout, while others had a short rostrum and still others a large and robust skull. Most had long, conical and pointed teeth, with fine grooves running from the base of the crown all the way to the point. These teeth could be small and delicate (as generally in "true" plesiosaurs) or large and robust (as in pliosaurs), according to whether they ate mollusks, fish, or other marine vertebrates. Although it is likely most species were active predators, some may have been scavengers. Dentition might seem to have been fairly uniform among this group, but there are nevertheless exceptions, because some forms (e.g., *Kaiwhekea*) had very small teeth crowded closely together and others (e.g., *Pliosaurus*) had large teeth that were triangular in cross section.

Animals resembling the pliosaurs—for example, rhomaleosaurids such as *Rhomaleosaurus* (7 meters)—are known beginning from the Lower Jurassic (fig. 2.31, top). The pliosaurids (plesiosaurs of the group Pliosauroidea) swam the seas mostly from the Middle Jurassic to the Late Cretaceous. The pliosaurids of the Early Jurassic were usually of modest dimensions, but starting in the Middle Jurassic they grew tremendously, giving us some of the most terrifying mega-predators: *Liopleurodon* (fig. 5.20, pp. 158–59),

▲ Fig. 2.32. *Occitanosaurus*, an elasmosaurid plesiosaur from the Lower Jurassic in the Aveyron (France).

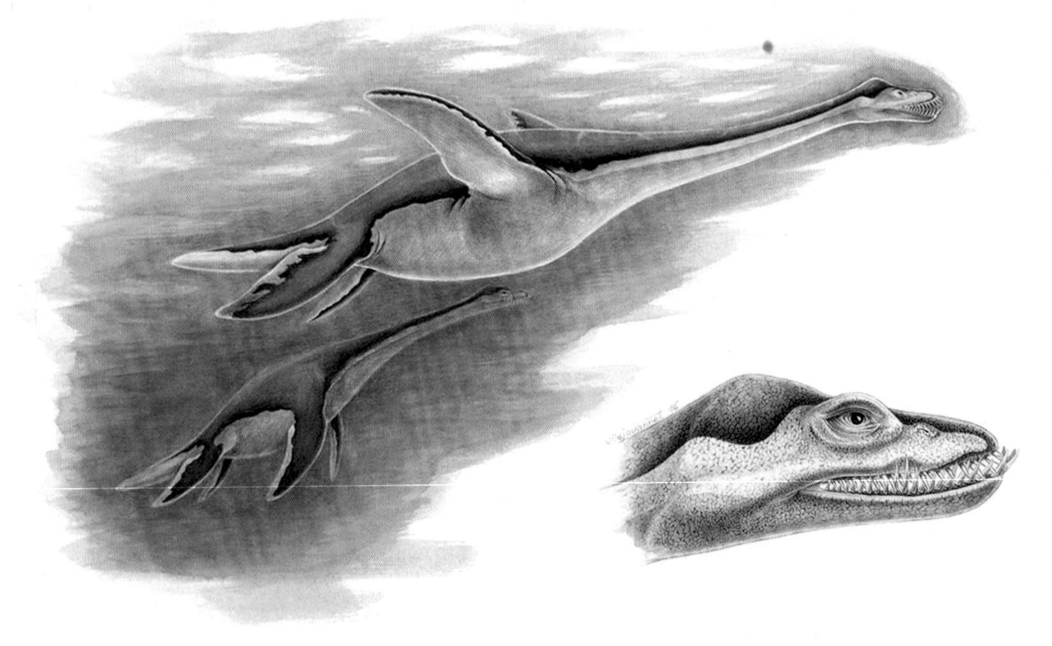

► Fig. 2.33. *Cryptoclidus*, a cryptoclidid plesiosaur from the Middle Jurassic in Europe, and its skeleton preserved at the Muséum national d'Histoire naturelle (Paris, France).

Pliosaurus, and *Kronosaurus* (fig. 6.3, pp. 165–66). The largest specimen of *Liopleurodon*, from the Middle Jurassic in Great Britain, exhibits a skull 1.5 meters long. The pliosaurids of the Late Jurassic and the Cretaceous grew even larger, and some, such as *Kronosaurus* and *Pliosaurus*, had a skull more than 2 meters long. A pliosaurid mandible found in England actually reached the 3-meter mark! These leviathans all had a powerful, long triangular muzzle, equipped with conical, sharp, and slightly curved teeth. They must have fed on fish of all sizes, as well as on other marine reptiles. In their short neck they had at most about twenty vertebrae; *Kronosaurus*, for instance, only had twelve.

The plesiosaurs proper (or "true" plesiosaurs) belong to Plesiosauroidea, which divides into lineages with very different characteristics: the plesiosaurids, the cryptoclidids, and the elasmosaurids (fig. 2.30). The plesiosaurids were creatures of relatively modest dimensions, known mostly from the Lower Jurassic in Europe. Their neck was long in proportion to their body and contained more than thirty vertebrae. It is within this group that we find *Plesiosaurus*, the very first genus of Plesiosauria to have been discovered and described (see

Plesiosaurus

Plesiosaurus was the first plesiosaur to be discovered, and by no less than noted paleontologist Mary Anning (see pp. 40–41). It was also the very first plesiosaur to be scientifically studied, described, and named. *Plesiosaurus* was an animal of modest size (3 to 4 meters in length) that lived around 190 million years ago, during the Early Jurassic.

Several relatively complete specimens found on the southern coast of England, at Lyme Regis, in Dorset (fig. 2.18), helped paleontologists gain more knowledge about *Plesiosaurus*'s anatomy. Its long neck contained about forty cervical vertebrae. Its tail was short, and its swimming paddles well developed, which indicates that it moved not by undulating its body but by use of its paddles. Its small skull, as well as the fifty delicate, sharp conical teeth that lined its jaws, probably did not allow it to hunt large prey. It must have subsisted on fish and invertebrates.

▲ Fig. 2.34. *Plesiosaurus*, a plesiosaurid plesiosaur from the Lower Jurassic in England.

Elasmosaurus

Elasmosaurus was discovered in sedimentary layers dating to about 80 million years ago in Kansas (in the central United States) and was described by American paleontologist Edward Drinker Cope (1840–1897) in 1868. This animal's anatomy—and notably its exceptionally long neck—were so troubling that young Cope made a monumental error when describing it: he placed its neck where the tail should have been, thinking that, like some dinosaurs, this animal had a long tail and a short neck. This error was the basis of a memorable quarrel with another American paleontologist, Othniel Charles Marsh (1831–1899), who publicly humiliated Cope by placing the neck where it belonged. The mutual hatred that animated these two men for the rest of their lives as a result of this quarrel gave rise to an extraordinary scientific competition between them (later called "The Bone Wars"), in which each raced to find more fossils than the other. Thanks to the Bone Wars, dozens of species were discovered, and hundreds of specimens of both dinosaurs and marine reptiles found, in just a couple of decades. This earned both Cope and Marsh enduring fame.

Because of this history, and because of *Elasmosaurus*'s immeasurably long neck, with more than seventy cervical vertebrae, it is the most famous genus of plesiosaur today. Its celebrity notwithstanding, the anatomy of *Elasmosaurus* is not well known, because remains of its limbs and part of its skull have never been found. Fortunately, the discovery of closely related plesiosaurs has allowed us to fill in some of those gaps. *Elasmosaurus* must have had a slender triangular skull equipped with needlelike teeth that decreased in size toward the back of the jaw. The upper and lower rows of teeth must have meshed perfectly once the animal closed its mouth. Its front limbs must have been longer than its rear limbs, and its tail must have been relatively short. It was a gigantic animal, just under 10 meters long.

▼ Fig. 2.35. *Elasmosaurus*, an elasmosaurid plesiosaur from the Upper Cretaceous in North America.

"*Plesiosaurus*," p. 55). Certain plesiosaurids exhibit a remarkable elongation of the neck; some, like *Occitanosaurus* (fig. 2.32), had a neck twice the length of their trunk.

Cryptoclidids were of modest dimensions too. *Cryptoclidus* (fig. 2.33) could reach up to 3 meters in length, with a long neck composed of thirty to forty vertebrae. These plesiosauroids are known mainly from fossils dating from the Middle Jurassic in England.

The elasmosaurids, known only from the Cretaceous, with their extremely long neck, were of a much more impressive size. The greatest numbers of cervical (neck) vertebrae belong to *Elasmosaurus* (which had seventy-two) (fig. 2.35) and *Albertonectes* (seventy-six). The latter sported a neck that was roughly 7 meters long! *Mauisaurus* had sixty-eight cervical vertebrae, *Hydralmosaurus* sixty-three, and *Styxosaurus* (which had a neck almost three times the length of its trunk) sixty-two. Despite the unknown ecological function of such an extreme elongation, elasmosaurids came in a wide variety, and they have been found at all latitudes, in sediments dating right up to the very end of the Cretaceous. The greatest number of specimens have been found in the United States.

The leptocleidids and the polycotylids are two other groups within Plesiosauria, but their precise fit is the subject of controversy. Some paleontologists classify them within Pliosauroidea; others, within Plesiosauroidea. They do exhibit a confusing mix of features—their skull was long, like pliosaurids', but slender, like plesiosaurids'. The polycotylids, such as *Dolichorhynchops and Thililua*, had a very long rostrum, and their neck, although relatively short, contained up to thirty vertebrae. These animals could reach considerable dimensions—*Pahasapasaurus* (fig. 2.36), for instance, was almost 6 meters long. A polycotylid fossil about 80 million years old, found in the United States, has provided proof of their viviparous nature, in the form of a juvenile preserved within the abdomen of an adult. The bones of the juvenile seem not to have been subjected to any degradation due to predation, so they

must be those of an embryo. It is likely that all species in Plesiosauria gave birth to live young, especially because a viviparous nature was already present in their close relatives, the pachypleurosaurs, back in the Triassic (see p. 48).

▲ Fig. 2.36. *Pahasapasaurus*, a polycotylid plesiosaur from the Upper Cretaceous in South Dakota (United States), shown giving birth. Its name comes from a Sioux (Lakota dialect) word meaning "black hills," referring to the name of the mountain chain where the fossil was found (the Black Hills).

The Fall of the Sauropterygians

Sauropterygians experienced an evolutionary success without equal among marine reptiles, but most went extinct before the end of the Mesozoic. Even though many hypotheses have been advanced, the causes of sauropterygians' slow disappearance are not completely clear. The only group that endured until the very end of the Cretaceous was the plesiosaurs proper, represented by the elasmosaurids.

At the beginning of the Cretaceous, plesiosaurs in the wide sense were still extremely diverse, with four large groups: the elasmosaurids, the polycotylids, the leptocleidids, and the pliosaurids. During the Late Cretaceous, the leptocleidids and the pliosaurids disappeared, and the polycotylids followed them. Did the plesiosaurs become extinct gradually? It is possible that the mosasaur boom (see p. 90) during the Late Cretaceous entailed sharp competition and a battle for survival that plesiosaurs lost over time. But during the Maastrichtian (the last age of the Cretaceous, 70–66 Ma), elasmosaurids still existed worldwide, in a wide variety of species—this apparent continued success does not fit such a pattern. Perhaps, of all the plesiosaurs, elasmosaurids' peculiar characteristics allowed them to avoid direct competition with the mosasaurs. We do not know. What we do know is that the consequences of the environmental upheavals that occurred at the very end of the Cretaceous were fatal to them.

The Hupehsuchians: An Exclusive Group

The hupehsuchians were both a short-lived and a geographically very limited group: they are known exclusively from around 248 million years ago, in the Lower Triassic (the Spathian stage), and from two very close counties in Hubei province—from which their name derives—in eastern central China (see fig. 4.8, p. 122).

Although we have known about hupehsuchians since the 1950s, only very recently, thanks to the unearthing of new fossils, has a clear picture of their phylogenetic relationships been established. They were actually first considered sauropterygians, then "**thecodonts**" within the archosaur group, and then a separate order of marine reptiles, closer to the ichthyosaurs. The new fossils have allowed researchers to identify hupehsuchians as diapsids closely related to the ichthyosaurs, with whom they form the clade Ichthyosauromorpha.

The hupehsuchians comprise five genera, all of them monospecific (one species only) and often based on very few specimens—sometimes only one. The diversity of this isolated group, all the specimens of which have come from the same deposit (or from deposits in very close proximity) and the same stratigraphic level, may be overestimated, even though the differences among specimens justify assigning them different genera.

Hupehsuchians' anatomy (fig. 2.37) is surprising. They had a massive body; a long muzzle; a long tail; and, generally, large, fan-shaped swimming paddles. They can be divided into three size categories: *Nanchangosaurus* and *Eohupehsuchus* were the smallest (roughly 40 centimeters long), *Hupehsuchus* and *Eretmorhipis* were roughly 1 meter long, and *Parahupehsuchus* could attain a length of almost 2 meters. Their skull exhibits two temporal fossae, indicating

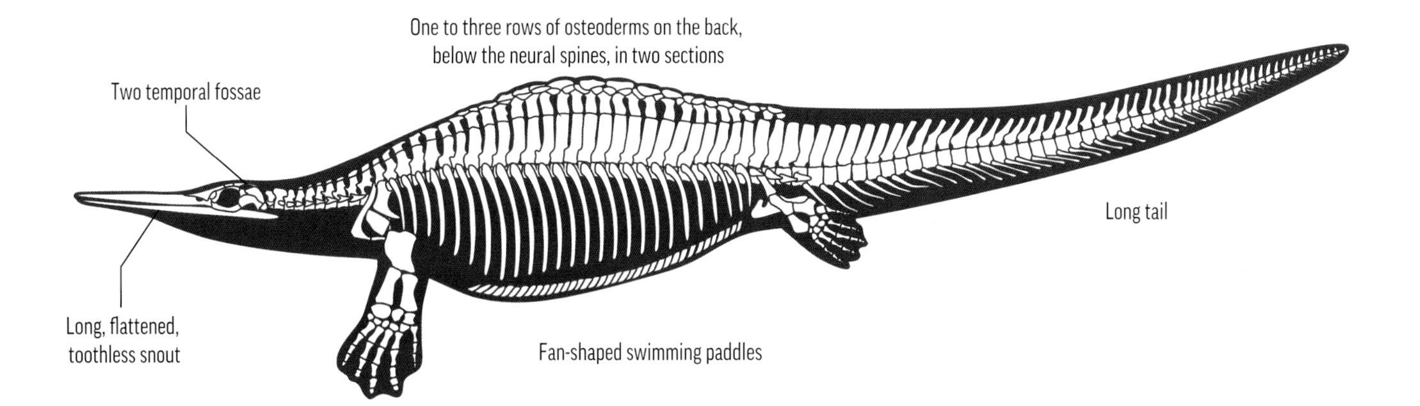

▲ Fig. 2.37. The very peculiar features of hupehsuchians, shown in *Hupehsuchus*, from the Lower Triassic in China.

that they belong to the diapsids, and a very long, slim muzzle, formed almost exclusively by the premaxillary bone. Their muzzle was laterally compressed, somewhat like the beak of a wading bird, and completely toothless. Their neck was generally longer than that of other vertebrates, containing nine or ten cervical vertebrae, with the exception of *Eohupehsuchus*'s six.

The hupehsuchians' body formed a unique "bony tube" (fig. 2.38) that was more pronounced in some genera and that extended from the pectoral girdle to the pelvic girdle. It was formed by the overlap of the ribs and the gastralia, both components being pachyostotic—similar to a turtle's shell (see p. 76), except that the girdles remained on the outside. This means that the ribs and gastralia coming together to form a protective shell, either externally (in turtles) or internally (in hupehsuchians), has happened at least twice in the history of the vertebrates. The bony tube extended along the entirety of the body in *Parahupehsuchus* but was limited to the pectoral region in the case of *Eretmorhipis*. In the case of *Hupehsuchus*, a faint space still existed between the ribs. This tube must have made each hupehsuchian's body considerably rigid, meaning that in order to swim it must have depended on the undulation of its powerful tail, like a crocodile.

Hupehsuchians' neural spines were all very short and, uniquely, composed of two separate parts. These spines were also sculpted toward the tip, which allows the possibility that they protruded from the skin. Another remarkable characteristic of the hupehsuchians was the presence of three layers of dermal plates (except in the case of *Nanchangosaurus*, which had only one layer) above the neural spines of the trunk and the base of the tail. These osteoderms were stacked in an alternating arrangement, like tiles, to leave no chinks in the animal's armor. In *Eohupehsuchus* the plates in all three rows were very small, whereas in the case of *Eretmorhipis*, *Hupehsuchus*, and *Parahupehsuchus* the ones in the third row were large, each covering several vertebrae—up to four in the case of *Eretmorhipis*!

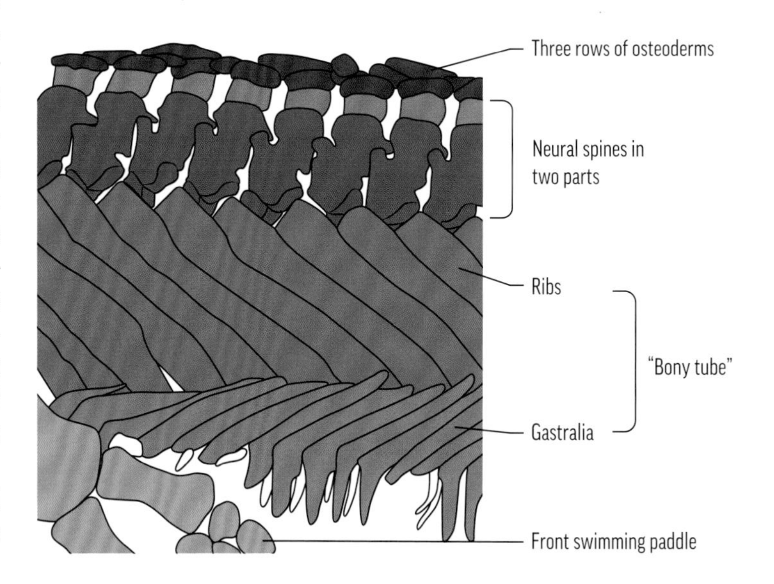

▼ Fig. 2.38. A diagram showing the "bony tube," formed by the conjoined and overlapping ribs and gastralia; the neural spines in two parts (two shades of gray); and the three rows of osteoderms in the hupehsuchian *Parahupehsuchus*. All this, which formed a robust bony armor, must have provided the animal with good protection from predators.

▲ Fig. 2.39. *Hupehsuchus*, a hupehsuchian from the Lower Triassic in China.

With the exception of *Nanchangosaurus*, the hupehsuchians' swimming paddles were well developed. In *Parahupehsuchus* they were slender; in *Hupehsuchus* (fig. 2.39) and *Eretmorhipis* they were fan-shaped. Hupehsuchians with wide and short paddles also exhibit hyperdactyly. This hyperdactyly is different from that found in ichthyosaurs but reminiscent of the first tetrapods of the Devonian, such as *Ichthyostega* and *Acanthostega*: another example of convergence (see "Convergence," pp. 198–99).

Although the deposits in which hupehsuchians have been found are otherwise rich in fossils, they are characterized by both a surprising absence of fish and a plethora of small marine reptiles about 20 centimeters to 1 meter long (comprising the ichthyosaur *Chaohusaurus* and the sauropterygians *Hanosaurus* and *Keichousaurus* in addition to five types of hupehsuchians). These creatures must all have been potential prey of the ecosystem's assumed mega-predator, an undiscovered sauropterygian 3–4 meters long. All of this shows that the trophic networks were fairly well established about 4 million years after the catastrophic Permian/Triassic crisis. The joint presence in hupehsuchians of a "bony tube" and of neural spines covered by up to three rows of osteoderms represents a unique and sophisticated mode of protection against predators. These defensive features must have made hupehsuchians heavy and somewhat inflexible. Considering their long free tail and powerful swimming paddles as well, this suggests that the area they lived in was a shallow sea, in which they typically kept to the bottom. Their slender, flat, and toothless muzzle seems to point to their being **filter feeders**, somewhat in the manner of modern beaked whales. The differences between these animals in terms of size, the length of the "bony tube," dermal armor, and paddle shape probably reflect a partitioning of the ecological niches among them so as to best exploit the food resources of the region they shared.

The Thalattosaurs: Enigmatic Reptiles

The name "thalattosaur" comes from the Greek *thálassa*, "sea," and *sauros*, "lizard," and therefore means "sea lizard." It refers to a group of Triassic marine reptiles that were widely distributed in the northern hemisphere, from Alaska to China, by way of Canada and Europe. Although this group's representatives were well distributed geographically (fig. 2.40), as well as relatively diverse, their fossils are fairly rare. Thalattosaurs roughly resembled lizards and were between 1 and 4 meters in length.

The oldest known thalattosaurs are from the Lower Triassic in North America. The great marine transgression of the Middle Triassic would facilitate their diversification and dispersal throughout the Tethys and the eastern portion of Panthalassa. Their transoceanic distribution is all the more remarkable since these animals were of only medium size and were not completely adapted to a pelagic way of life (in the open sea). Their disappearance at the end of the Triassic is unexplained. Did competition with the ichthyosaurs (see p. 30), which were constantly growing in number and above all were increasingly specialized, help vanquish the thalattosaurs, animals of a rather generalist nature? Because they are so rare in the fossil record to begin with, we have little to go on.

Thalattosaurs' anatomy indicates that, although they were adapted to an aquatic way of life, they were probably amphibious and therefore capable of returning to land for both warmth and purposes of reproduction. Their limbs were short and probably ended in flippers, but they retained some impressive front claws, which must have allowed these animals to move about on the shore, as well as to resist swells while they rested on the rocks, like today's marine iguanas. In illustrations, their tail is often represented as raised (in other words, stretched dorso-ventrally), like salamanders' and newts', but there is no reason to believe that it wasn't slender and whip-shaped like that of iguanas (see fig. 1.2, p. 9). Their tail was exceptionally long, even for a reptile—up to twice the length of their trunk! Coupled with the undulation of the entire body, that tail probably ensured efficient propulsion in water. Their snout was generally long and slender. Their nostrils, like those of many animals that have returned to a marine way of life, were situated behind the snout, near the eyes. Also, like those of many aquatic animals (but not exclusive to them), their eyes were protected by sclerotic rings (see "Diving and Underwater Vision," p. 150–51).

Thalattosaurs' exact position within the class Reptilia remains extremely fluid and

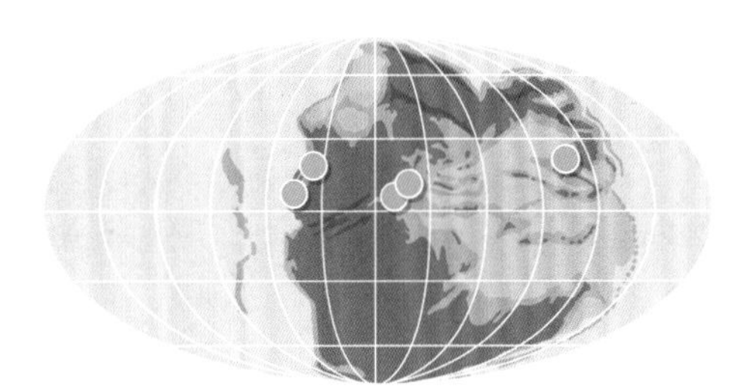

Early and Middle Triassic

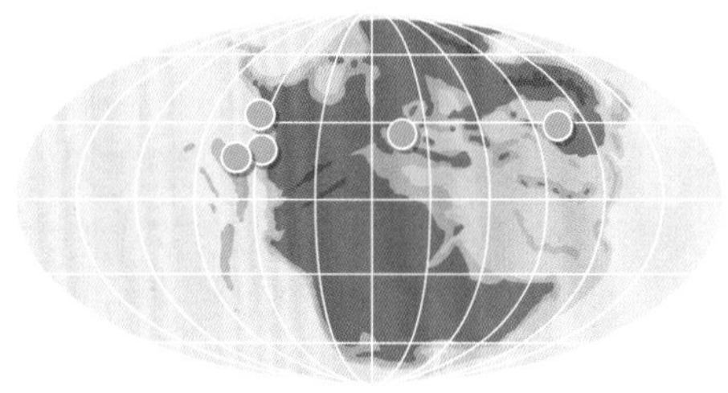

Late Triassic

◄ Fig. 2.40. The distribution of the thalattosaurs over the course of the Triassic.

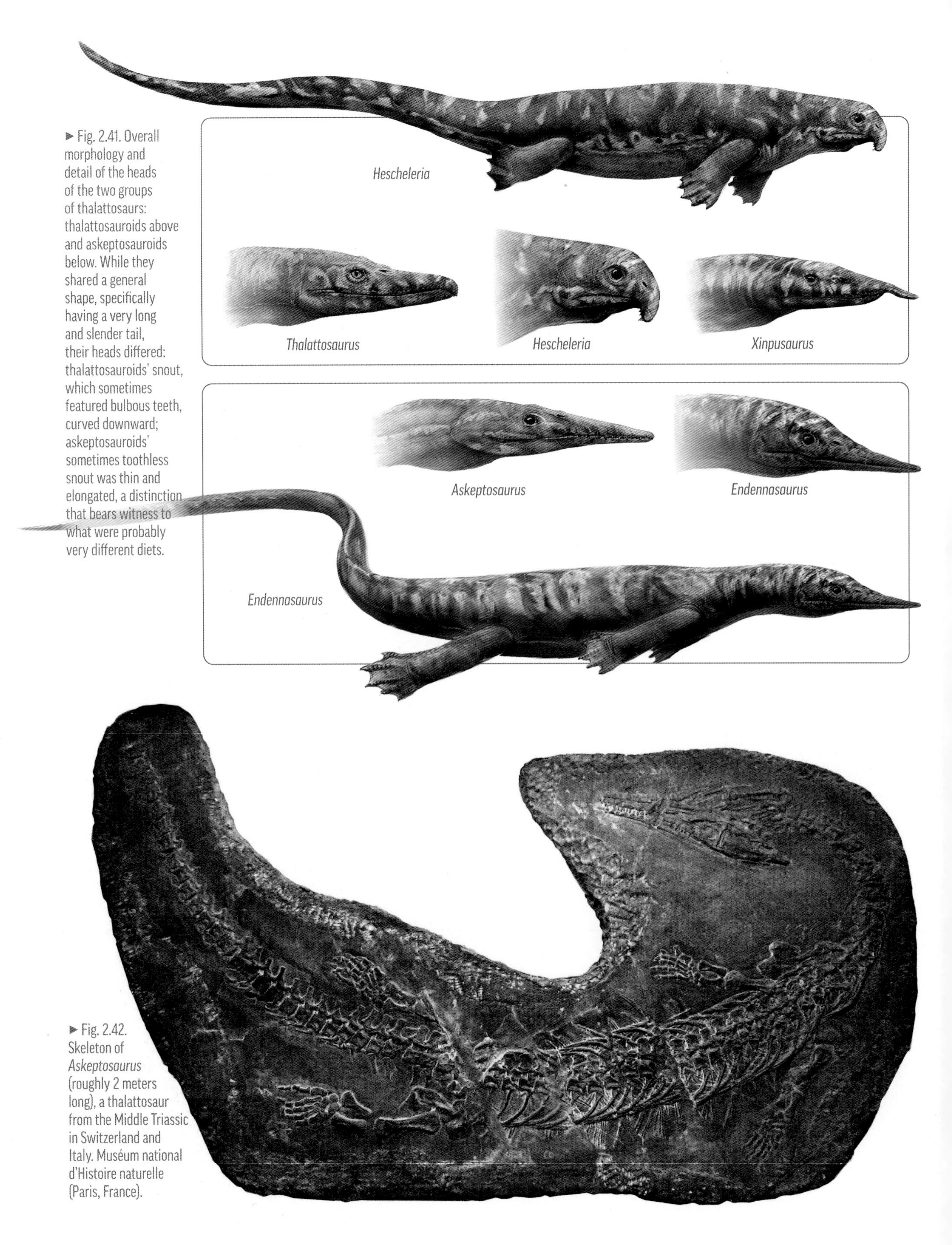

► Fig. 2.41. Overall morphology and detail of the heads of the two groups of thalattosaurs: thalattosauroids above and askeptosauroids below. While they shared a general shape, specifically having a very long and slender tail, their heads differed: thalattosauroids' snout, which sometimes featured bulbous teeth, curved downward; askeptosauroids' sometimes toothless snout was thin and elongated, a distinction that bears witness to what were probably very different diets.

► Fig. 2.42. Skeleton of *Askeptosaurus* (roughly 2 meters long), a thalattosaur from the Middle Triassic in Switzerland and Italy. Muséum national d'Histoire naturelle (Paris, France).

controversial. While they are most often considered diapsids (see p. 25), they have at different times been regarded as close to the sauropterygians, the lepidosauromorphs, and the archosauromorphs. Whatever the case might be, the thalattosaurs themselves can be clearly divided into two groups: the thalattosauroids and the askeptosauroids (fig. 2.41).

In most thalattosauroids, the end of the snout curved downward, and the short, massive round teeth of some species seem to indicate a diet of tough prey. Their curved snout might have helped them manipulate shelled creatures and other mollusks or aided their foraging on the seafloor. Thalattosauroids occupied numerous ecological niches and were perfectly adapted to life in shallow coastal waters and reef systems.

Askeptosauroids, such as *Askeptosaurus* (figs. 2.42 and 2.44), with their short pointy teeth, were probably opportunistic surface predators that ate fish and any prey within their reach. Some, like *Endennasaurus* (fig. 2.41), were practically toothless.

Hescheleria

Hescheleria was a small, very particular thalattosauroid discovered in the 1930s in Middle Triassic deposits (from 247 to 235 Ma) at Monte San Giorgio, on the border between Switzerland and Italy (see chapter 4, p. 113). Although its overall appearance, like that of other thalattosaurs, was that of a lizard, with a very long tail and short limbs, its snout was strange. Its lower jaw was short, and the end of its upper jaw was turned downward, almost vertically, covering the front part of the mandible when the creature's mouth was closed.

At the time of its original description in 1936, *Hescheleria* was thought to most likely have fed on mollusks, the curved part of the skull helping hold the shells while the lower teeth crushed them. Yet some experts expressed doubts, since some of the teeth in the mandible closed upon a diastema (a section devoid of any teeth) in the upper jaw, which obviously would have made crushing difficult. Therefore the function of this snout remains somewhat a mystery.

▲ Fig. 2.43. *Hescheleria* (roughly 1 meter long), a thalattosaur from the Middle Triassic in Switzerland.

Askeptosaurus

Askeptosaurus was an askeptosauroid about 2.5 meters long, found in both Italy and Switzerland. Its tail accounted for about half of the animal's entire length and must have allowed it to swim using undulation. Its limbs, short and relatively weak, probably allowed it to move about on land for only short distances, for purposes such as laying eggs.

Askeptosaurus means "unsuspected lizard," and its discovery in 1925 was indeed fortuitous. Hungarian paleontologist Franz Nopcsa von Felső-Szilvás was at the time studying the remains of a small ichthyosaur, *Mixosaurus*, which had been provided by the Natural History Museum in Milan, when he noticed some bones that did not belong to the ichthyosaur's skeleton. The baron proceeded to describe this new species, and several more specimens were found thereafter.

▲ Fig. 2.44. *Askeptosaurus*, a thalattosaur from the Middle Triassic in Switzerland and Italy.

The Crocodylomorphs: Variations on a Crocodile

Like dinosaurs and pterosaurs, crocodylomorphs are archosaurs (figs. 2.3 and 2.4). In today's natural world they are represented only by the crocodiles (or crocodilians). But that was not always the case. During the Mesozoic, crocodylomorphs came in an impressive variety: they occupied many ecological niches, and their physical attributes differed much more substantially than they do today.

Amphibious Crocodiles? Not Necessarily

All of today's crocodylomorphs, whether we are talking about crocodiles, caimans, alligators, or gavials, are amphibious, and only *Crocodylus porosus* (see chapter 1, p. 10), the saltwater crocodile, occupies the marine environment in a recurring fashion. Their Mesozoic cousins were a much more mixed

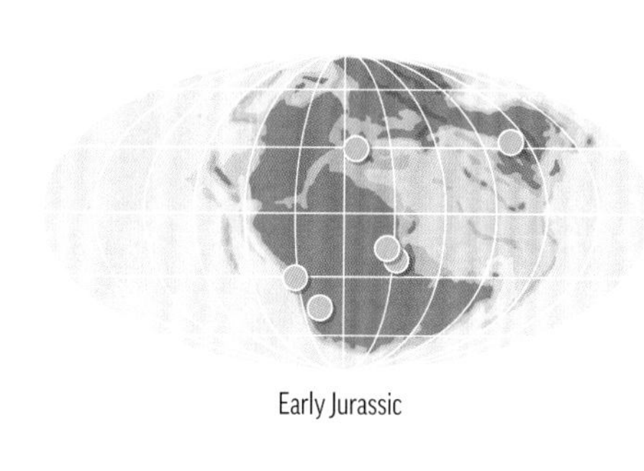

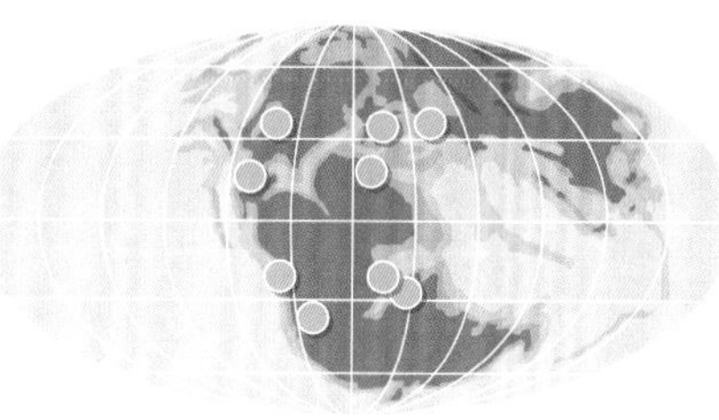

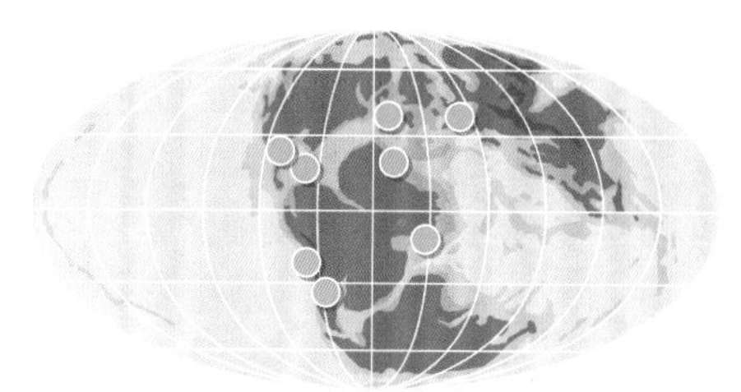

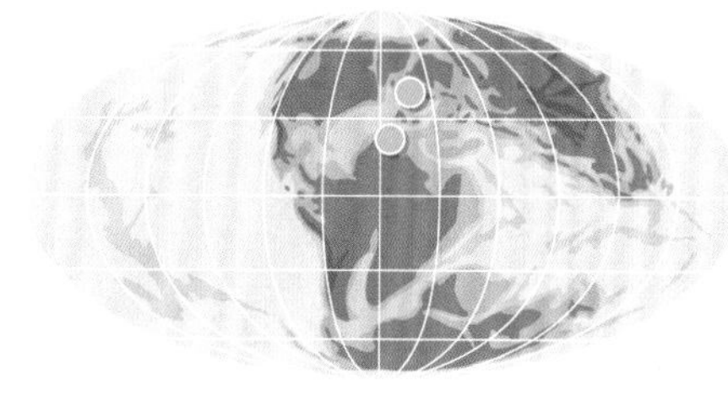

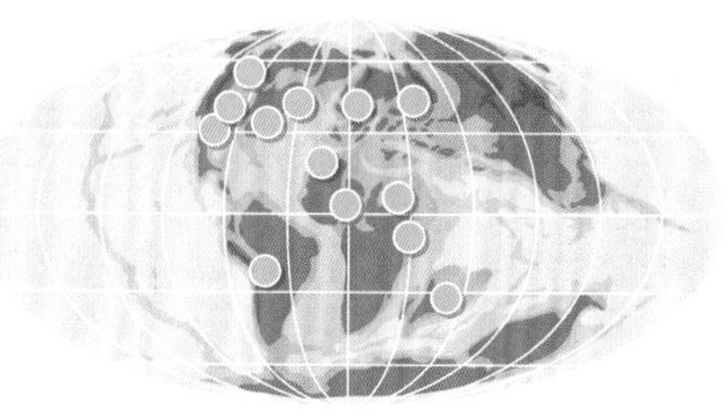

◄ Fig. 2.45. The distribution of marine crocodyliforms during the Mesozoic.

group. Numerous forms were completely terrestrial, such as *Bretesuchus* and *Sebecus*, the skull of which more closely resembles that of a tyrannosaur than that of other crocodilians. Some notosuchians, such as *Simosuchus*, were even herbivorous. The most basal (least evolved) crocodylomorphs, such as the protosuchians (fig. 2.46), were terrestrial, and it therefore seems that creatures in this group only subsequently began to adapt to aquatic environments.

Aquatic forms are numerous among the fossil crocodylomorphs, and it is sometimes difficult to distinguish the forms that were completely marine from those that only occasionally visited the ocean or only frequented brackish estuaries or mangrove swamps. Bear in mind that, like today's saltwater crocodile, some fossil species must have spent the first years of their lives in fresh water. We know this because the largest specimens of those species are found in areas that correspond to the sea and smaller ones are found in areas corresponding to rivers. In addition, river currents sometimes transport the corpses of non-marine aquatic animals to the sea, where they can become fossilized in marine sediments. Rather than discuss these groups with an uncertain lifestyle, we will consider only the groups that can frequently be found in marine sediments.

◄ Fig. 2.46. *Gobiosuchus*, a terrestrial protosuchian from the Upper Cretaceous in the Gobi Desert, Mongolia.

▲ Fig. 2.47. *Teleosaurus*, a thalattosuchian from the Middle Jurassic in Normandy (France), showing the rows of osteoderms.

Some True "Marines": The Thalattosuchians

As their name reveals, the thalattosuchians (from the Greek *thálassa*, "sea," and *suchos*, "crocodile") were, with few exceptions, exclusively marine crocodylomorphs. The oldest traces of representatives of this group go back to the Sinemurian stage (Lower Jurassic sediments roughly 195 million years old). Thalattosuchians' geographic origin is not clear, since the most ancient of their fossils come from very distant locations: South America, India, and France (fig. 2.45). Thalattosuchians from the end of the Early Jurassic and later are known mostly from Europe and the Neuquén Basin in Argentina.

The thalattosuchians are traditionally divided into two large groups: the teleosauroids and the metriorhynchoids. The teleosauroids, the more ancient of the two, exhibit much less pronounced adaptations to the aquatic environment, with an overall anatomy not much different from that of other amphibious crocodylomorphs (fig. 2.47). They were equipped with dorsal and ventral shields, made of bony plates formed in the skin (osteoderms). Propulsion was supplied by a long, supple, muscular tail. The forelimbs were certainly a little small, but they still allowed for movement on land both to lay eggs and to warm up in the sun. All teleosauroids were equipped with a long rostrum (fig. 2.49), which suggests a diet mostly of fish, or sometimes tortoises for those forms with more robust teeth. Given their aptitude for swimming and the dorsal position of their orbits in the skull (their eyes were on top of their heads), most likely these animals evolved in coastal environments, avoiding the open ocean and preferring to hunt by ambush from below.

The origin of teleosauroids, and therefore of thalattosuchians, is still much debated (fig. 2.50). For a long time, paleontologists were guided by the fact that thalattosuchians' skull, especially the palate and the bones around the ear, was very primitive for crocodylomorphs. However, most recent phylogenetic research places them closer to other, more evolved, long-snouted saltwater crocodylomorphs: the pholidosaurids and the dyrosaurids. It is possible that this grouping is actually an artifact: these groups' similarly long muzzles might simply be a result of shared ways of life and shared dietary regimens, rather than a result of close evolutionary kinship (i.e., an example of convergence; see "Convergence," p. 198–99). Only the discovery of really ancient fossils resembling the first thalattosuchians but still retaining some ancestral characteristics would allow experts to decide. Unfortunately, the oldest known significant thalattosuchian forerunner, *Peipehsuchus*, from about 180 million years ago (the Toarcian) in China, was already a very long-snouted form exhibiting all the

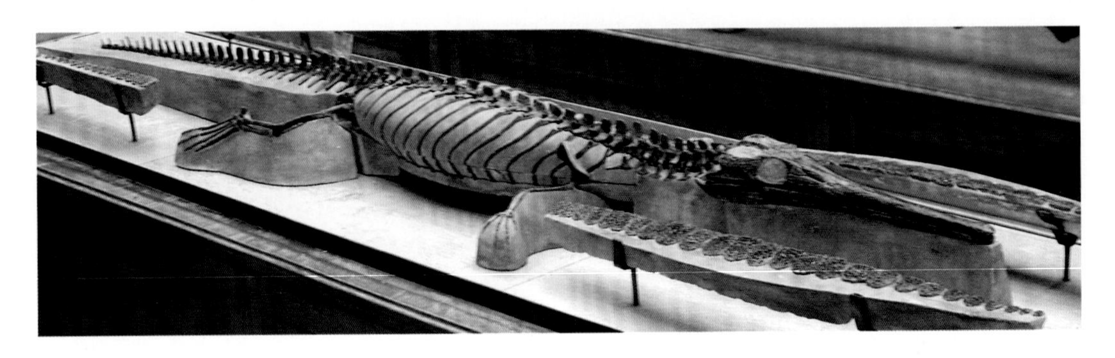

► Fig. 2.48. *Pelagosaurus*, a thalattosuchian from the Lower Jurassic in Calvados (France), Muséum national d'Histoire naturelle (Paris).

characteristics of the thalattosuchians and of the teleosauroids specifically.

The oldest fossil metriorhynchoids all come from the Toarcian, as *Peipehsuchus* does, and have been found in Europe. They exhibit increasingly marked skeletal modifications that point to a much more exclusive aquatic life: the forelimbs are short, and the fingers of both hands and feet are flat but show no hyperphalangy or hyperdactyly (compare "From Legs to Swimming Paddles," p. 39); in addition, their skeleton was light, composed of spongy and not very compact bones, and they lacked a protective shield, either dorsal or ventral. They were equipped with a hypocercal tail—one with a smaller upper lobe than in ichthyosaurs, the presence of which nonetheless demonstrates a high degree of adaptation to the aquatic environment and to rapid swimming. As in the case of the ichthyosaurs, the caudal vertebrae supported the lower lobe (fig. 2.51).

Curiously, while the metriorhynchoids' forelimbs had shrunk, their hind limbs remained very long, comparable to the hind limbs of today's crocodilians. But the muscle insertion areas on the bones are not very marked, revealing a certain muscular weakness that would have limited the involvement of the hind limbs in locomotion. It must have been difficult, if not impossible, for metriorhynchoids to move about on land as modern crocodiles do. Perhaps these hind limbs served only directional or stabilizing purposes, like **hydrofoils** on ships and submarines.

Metriorhynchoids' skull presents numerous innovations that testify to a lifestyle that was more marine than that of the teleosauroids. First, certain fossils show evidence of an enlarged salt gland, which would have helped these creatures eliminate any excesses of salt

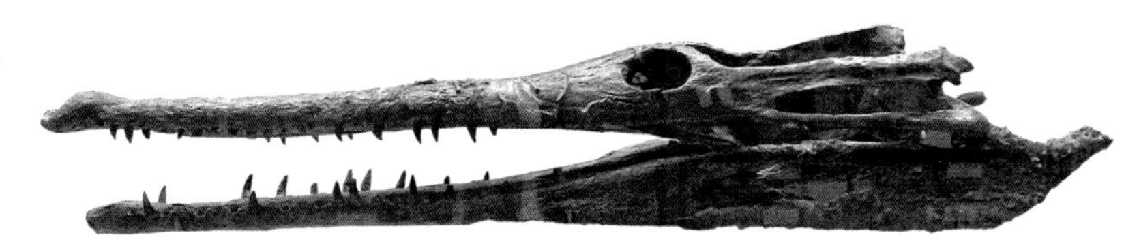

▲ Fig. 2.49. Skull of *Proexochokefalos*, a thalattosuchian from the Middle Jurassic in Calvados (France), Muséum national d'Histoire naturelle (Paris).

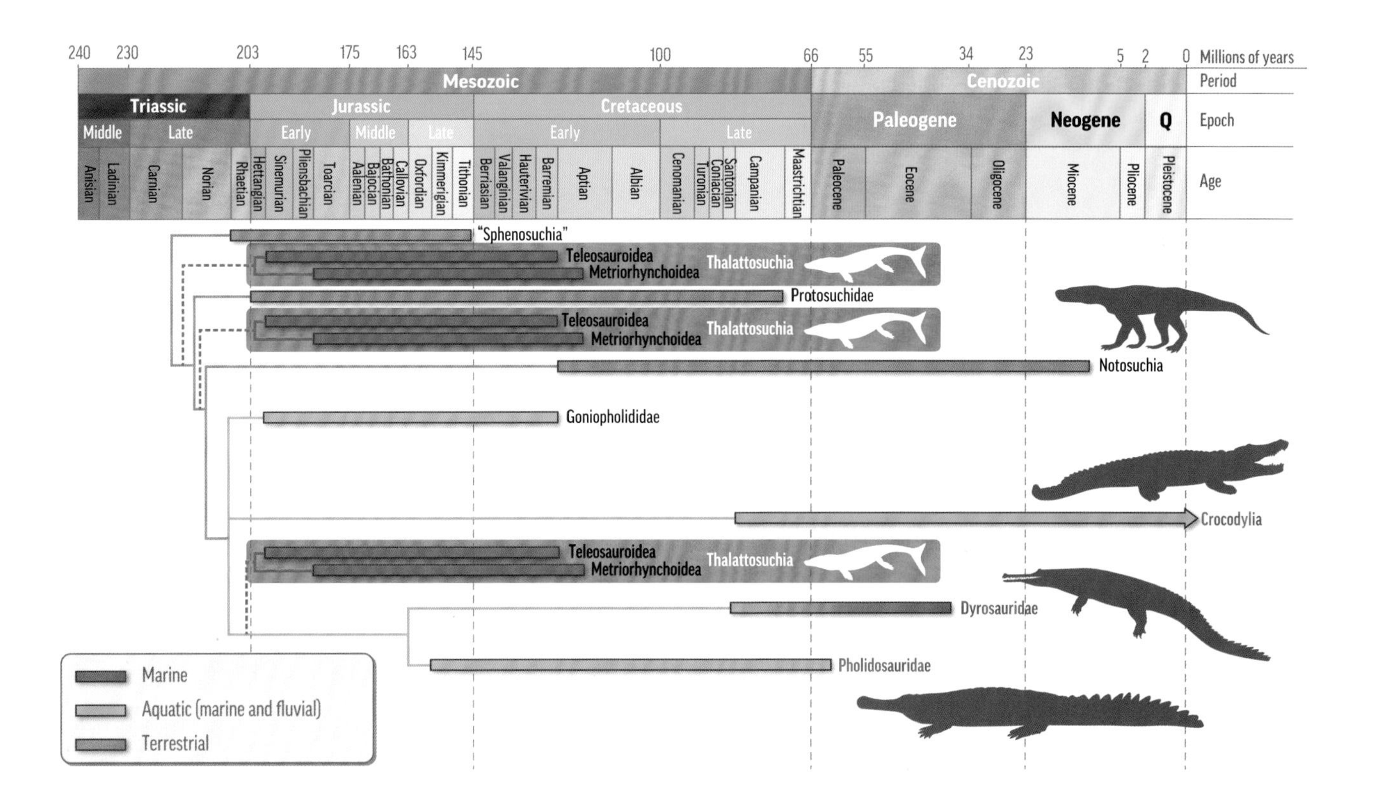

▼ Fig. 2.50. A simplified phylogeny of the crocodylomorphs, showing the groups mentioned in the text and the different possible positions for the thalattosuchians.

in the body that had not been excreted by the kidneys. Situated in a sort of bony visor in front of and below the eye sockets, this gland most likely was present in all metriorhynchoids. Moreover, the eyes were oriented no longer upward but laterally, which attests to a modification in predation behavior. The metriorhynchoids were therefore probably open-ocean hunters, capable of actively pursuing their prey (fig. 2.52).

No fossilized metriorhynchoid embryos have been found, but given these animals' great degree of adaptation to the aquatic environment, it should not be ruled out that they gave birth to live young. Nevertheless, metriorhynchoids may have used their hind limbs to move about on land to lay eggs. If so, given their feeble musculature, it must have presented a real challenge!

The metriorhynchoids seem to have favored different prey than the teleosauroids did. Metriorhynchoids had a significantly shorter muzzle and, overall, their teeth were more robust, which seems to reveal a more varied and opportunistic diet. *Dakosaurus* (see fig. 5.21, p. 161) pushed these characteristics to the extreme, with a very high, short, and squat skull containing teeth that were far apart but were massive and **ziphodont** (serrated). This morphology, a classic one in predatory dinosaurs, as well as in terrestrial crocodylomorphs, is rare among marine crocodylomorphs. Imagine the head of a *Tyrannosaurus* placed on a body capable of swimming extremely efficiently!

Thalattosuchians began to decline at the end of the Jurassic. For a long time, researchers supposed they became extinct shortly thereafter, at the beginning of the Early Cretaceous; however, recent reinterpretation of some fragmentary fossilized remains suggests that they died out about 20 million years later, at the end of the Early Cretaceous. What doomed them? For the time being, there seem

▼ Fig. 2.51. Skeleton of *Cricosaurus*, a thalattosuchian from the Upper Jurassic in Germany.

At almost 7 meters in length, *Plesiosuchus manselii* is the largest known metriorhynchoid. Discovered in Kimmeridgian–Tithonian sediments (from 155 to 150 Ma) in England, its short muzzle and serrated teeth made it a redoubtable predator, at the top of the food chain. It lived next to another mega-predator, *Dakosaurus maximus* (fig. 5.21, p. 161), a close cousin that was just as formidable; this

Plesiosuchus

cohabitation spurred researchers to formulate some questions about the sharing of food resources. Comparisons have therefore been made with today's cetaceans, and more specifically with the orcas of the North Atlantic, where two populations coexist: a population of small orcas, with teeth that exhibit much wear, and a population of larger ones, with teeth that are unworn. While the second group preys on large mammals, the first is much more generalist. The same differences in size and dental characteristics are present between *D. maximus* and *P. manselii*, thus suggesting an ecological division similar to that of today's North Atlantic orcas.

Plesiosuchus might have been the last thalattosuchian to die out, since the thalattosuchian specimen that dates to later than all the rest, around 115 million years ago (the Aptian), could be attributed to this genus.

▼ Fig. 2.52. *Plesiosuchus*, a thalattosuchian from the Upper Jurassic in England.

to be no clear answers. Reconstructing such ancient histories is no simple matter. ... In this case, the transition from the Jurassic to the Cretaceous is defined by a very important biological crisis both on the continents and in the seas, in which the crocodylomorphs and, above all, the turtles seem to have suffered. The beginning of the Cretaceous was marked by a profound change in climate, as well as by low sea levels, which affected the faunas of the continental shelves. A corollary to this retreat of the oceans was diminished sedimentation, so marine fossil deposits from this epoch are relatively scarce. As a result, paleontologists lack knowledge of the marine faunas from this time. It is therefore possible that these faunas' diversity has been underestimated and that fossils of faunistic replacements have simply eluded researchers so far.

To Sea or Not to Sea: The Pholidosaurids Hesitate

Considerably less varied than the thalattosuchians, the pholidosaurids are represented by a little more than a dozen species, about half of which lived in rivers or estuaries. It is fairly difficult to determine to what extent certain species could venture into the marine environment. For example, some fossilized remains of *Pholidosaurus* around 145 million years old have been found in marine sediments from the Berriasian stage, but these may be examples of river-dwellers that were swept into the sea shortly after their demise.

Pholidosaurid fossils, particularly those that include parts of the body other than the head, are rare, a fact that does not help paleontologists arrive at definite conclusions on their degree of adaptation to the aquatic environment. The marine forms, such as *Terminonaris*, from around 90 million years ago (the Cenomanian–Turonian) in North America, and *Oceanosuchus* (fig. 2.53), from around 95 million years ago (the Cenomanian) in Normandy, all had very similar skulls. Their snouts were very long and their teeth very thin, which suggests a diet of fish. Their distinctiveness among crocodylomorphs is related to the end of their upper jaw: it formed a sort of "beak" in front of their mandible (fig. 2.53). The role of this beak is unclear. The rest of the typical pholidosaurid body, to the extent that it is known, does not seem much different overall from that of today's amphibious crocodilians. Pholidosau rids were equipped with limbs that allowed them to move about on land easily, and they were armored both top and bottom by highly developed structures of osteoderms. The marine forms were probably coastal and did not venture into the open ocean much.

Until 2020, pholidosaurids were known as having lived from the Bathonian age to the Turonian age (ca. 166 to 90 Ma). Then a specimen unearthed in Morocco was dated to around 60 million years ago (the Paleocene epoch), extending the evolutionary history of this group by about 30 million years and adding another crocodyliform to the list of reptiles that survived the Cretaceous/Paleogene crisis! In paleontology, as in all scientific domains, conclusions are never definitive.

Some Extraordinary Navigators: The Dyrosaurids

Dyrosaurid fossils are prevalently found in marine environments. From a phylogenetic point of view, dyrosaurids are very close to pholidosaurids (fig. 2.50). Their limbs were much more robust than those of today's crocodilians and equipped with much more powerful muscles. From this, we deduce that dyrosaurids undoubtedly spent time on dry land, where they must have been able to move with relative ease. Their tail is also worthy of mention. It was very prominent in terms of height, with neural and haernal spines more than twice as tall as those of modern crocodilians (fig. 2.54). Conclusion: the tail housed a musculature more than twice as massive as those of today's crocodilians. This must have enabled the dyrosaurids to be significantly more powerful swimmers than their present-day cousins, and even the latter manage to hoist almost their entire body above the water using only their tail for propulsion!

Most dyrosaurids had long and slender snouts, with very sharp teeth, particularly suited to catching fast and slippery fish. Some species (e.g., *Phosphatosaurus*), their relatively long snout notwithstanding, were equipped with impressively strong and massive teeth, comparable to Nile crocodiles', which points to a diet much wider than that of other dyrosaurids, which were essentially piscivorous.

Until recently, hypotheses about dyrosaurids' origin were relatively vague. Although one fragmentary fossil has been dated to about 95 million years ago (the Cenomanian), and even that is still in doubt, no other specimens definitively date to prior to the Campanian or even the Maastrichtian age (ca. 80 to 66 Ma). A "hole"—or a phantomatic lineage, as it has been called by paleontologists—of almost 100 million years thus exists between the age of the earliest dyrosaurids found (from the Upper Cretaceous) and their presumed evolutionary divergence from

▲ Fig. 2.53. *Oceanosuchus*, a pholidosaurid crocodile from the Upper Cretaceous in Normandy (France).

the pholidosaurids around 157 million years ago (in the Late Jurassic). Nonetheless, some phylogenies seem to establish a proximity, in terms of kinship, between some species from the Lower and mid-Cretaceous and the dyrosaurids, thus at least partially filling this hole.

Although data regarding dyrosaurids' variety during this time are spotty, fossils from the Maastrichtian show us the number of species was increasing. Nevertheless, it was the Paleocene epoch (66–56 Ma) that proved the real turning point, with a veritable explosion of diversity and a colonization of the marine environment.

Dyrosaurid fossils from the Maastrichtian are almost never found in the same deposits as mosasaur and plesiosaur fossils; they are found in those associated with fluviatile • environments (rivers and streams) instead. It seems that dyrosaurids profited from the extinction of other marine reptiles, notably the mosasaurs, at the end of the Mesozoic, diversifying to fill the vacated ecological niches. In their case, there was not a Cretaceous/Paleogene crisis but a Cretaceous/Paleogene opportunity! They spread to almost all continents, except Europe, where, bizarrely, they would never set foot. The African deposits in particular show us there was exceptional variety within this group until the mid-Eocene (45 Ma).

The dyrosaurids did not face extinction until well after the Cretaceous/Paleogene crisis, around 40 million years ago, a little before the end of the Eocene epoch. Their demise was perhaps due to changes in climate, a drop in sea level, the emergence of the cetaceans (see chapter 7, p. 192), or a combination of these factors.

The Survivors: The Crocodilians

Modern crocodilians are the only living representatives of the crocodylomorphs. They are

Dyrosaurus

Over 6 meters long, *Dyrosaurus* was a crocodylomorph of imposing size. Typically found in North Africa, it was discovered in the phosphate mines, in sediments from the Lower Eocene (ca. 50 Ma). Two species have been described: *D. phosphaticus* in Algeria and Tunisia, and *D. maghribensis* in Morocco. Its dimensions notwithstanding, this animal was a placid fisher, like today's Indian gavial, happy to eat fish with its long, thin snout and its narrow teeth. In the open-pit phosphate mines of the Khouribga region of Morocco (see chapter 6, p. 172), some Eocene layers are so rich in *Dyrosaurus* deposits that there is a nearly complete skeleton almost every 15 meters! This treasure trove, proof of the abundance of life in the region's seas during this period, is probably tied to frequent upwellings of plankton-rich waters from the depths of those Paleogene waters.

▲ Fig. 2.54. Skeleton and reconstruction of *Dyrosaurus*, a dyrosaurid crocodile from the Lower Eocene in North Africa. OCP Group (Khouribga, Morocco).

divided into three groups: the long-snouted, piscivorous gavialoids; the alligatoroids (alligators and caimans); and the crocodyloids (crocodiles and *Tomistoma*). Out of somewhat less than thirty species, only one, the estuarine crocodile (*Crocodylus porosus*), occupies the marine environment in a regular fashion. Its pretensions to being a marine animal, as we mentioned earlier (see chapter 1, p. 10), are fairly questionable, since it spends a great deal of its life in rivers or on dry land.

Crocodilians have been present in lakes, rivers, and marine environments ever since they first appeared, in the Cretaceous. They are thought to have arisen in the middle of the Cretaceous, but the oldest verified specimens are from around 80 million years ago (the Campanian), in North America and Europe. A full third of Mesozoic crocodilian species made significant inroads to the marine environment. This flexibility regarding their living environment probably proved a major asset, allowing them to traverse seas to conquer the rivers and lakes of other continents, and it certainly explains their wide distribution (fig. 2.45). It also helped reduce competition with the other large marine reptiles of the Mesozoic.

Gavialoids in particular were quite varied at the end of the Cretaceous and during the Paleocene epoch, between about 70 and 60 million years ago, and are principally known for their marine forms, which have been found over a vast area from North and South America to Europe and North Africa. Yet they exhibit no special adaptations to the marine environment that might differentiate them from crocodilians that inhabited rivers and lakes. Therefore, these gavialoids were probably coast-dwellers that only rarely frequented the open ocean.

Just like the dyrosaurs, crocodilians as a group made it through the Cretaceous/Paleogene crisis without a problem; certain species were present both before and after the crisis.

The Turtles: A Unique Anatomy among the Vertebrates

Most people imagine turtles as slow-moving, harmless creatures that can be cute pets. It's true that all modern turtles are toothless, but they *are* equipped with a very sharp horned beak. And although most are herbivores, some are redoubtable predators with a diet very different from salad leaves. In a painful reminder of differences that exist among turtles, in a lake in Germany in 2013, an alligator snapping turtle (*Macrochelys temminckii*) severed the Achilles tendon of an unfortunate child. So watch your fingers!

Turtles are the most remarkable of all reptiles in regard to their staying power: having emerged from the depths of the Triassic almost 250 million years ago, and having colonized practically all environments except for the coldest (fig. 2.55), they are still with us!

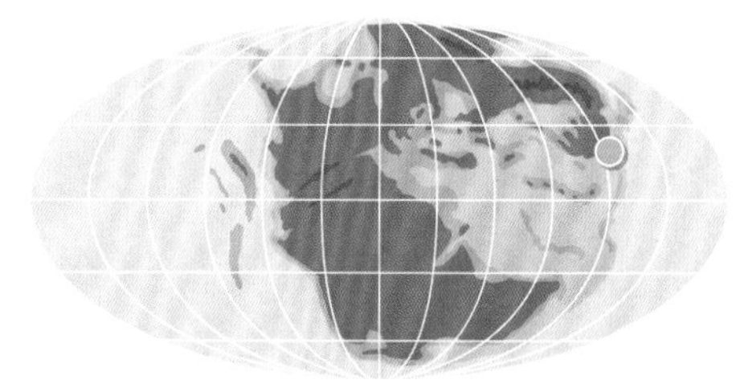

Late Triassic

Late Jurassic

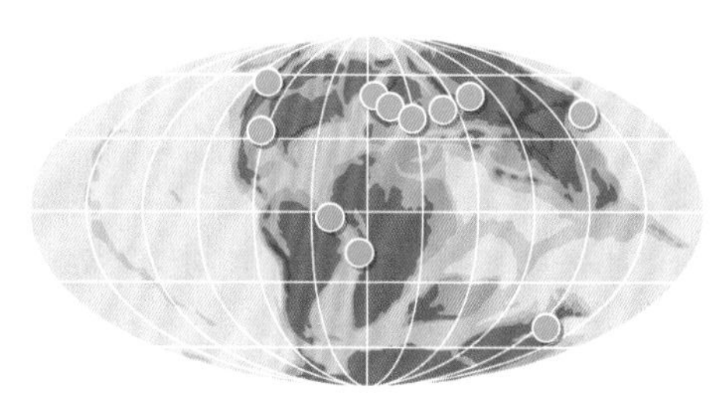

Mid-Cretaceous

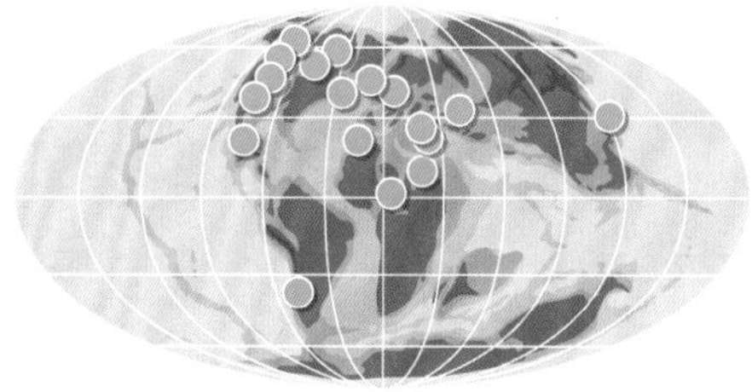

Late Cretaceous

◄ Fig. 2.55. The distribution of marine turtles during the Mesozoic.

In addition, turtles include the only marine reptiles besides plesiosaurs to use a swimming technique called underwater flight (see "Locomotion in Ichthyosaurs and Sauropterygians," pp. 148–49). Let us now focus on their completely atypical anatomy.

There Is Armored, and Then There Is Armored

Turtles' most defining attribute is their shell: the carapace on top and the plastron on the bottom, connected to each other by what is called the bridge. In the case of all other armored amniotes, such as certain dinosaurs (e.g., ankylosaurs and titanosaurs), crocodiles, placodonts (see fig. 2.24), and armadillos, the armor is composed of extra bones, called osteoderms, that develop in the skin to form an external shell that surrounds the skeleton. As we already stated when describing the placodonts, nothing even remotely similar occurs in the case of the turtles: their shell is formed by the ribs, the dorsal section of the vertebrae, and the clavicle, which expand to the point of coming together and frequently fusing. In other words, the shell is an integral part of the turtles' skeleton and is not the equivalent of the exoskeleton of armored animals we listed previously. The classic cartoon gag of a turtle completely exiting its shell is therefore impossible!

Uniquely among all vertebrates in the history of the world, turtles' pectoral and pelvic girdles are situated entirely inside the rib cage (fig. 2.56). So are the upper parts of the limbs. This peculiar anatomical configuration allows turtles (except for most aquatic varieties) to retract their head and their limbs into an armored structure: a perfect defense!

The question that naturally arises concerns the way the shell was formed as a whole and how the creatures' girdles made the transition from the outside of the rib cage to the inside. These anatomical uniquenesses—a veritable tour de force of evolutionary sleight of hand—have finally been better understood, thanks to some discoveries both in paleontology and in developmental biology in recent decades.

Finds of primitive fossils have shed light on the transition between chelonian ancestors without a shell to turtles with a shell. Until a relatively short while ago, the oldest known

▼ Fig. 2.56. A cutaway diagram of a turtle skeleton, demonstrating that the shell (carapace and plastron) is not external to the skeleton (as in the case of crocodiles) but is intrinsic, the result of a long evolutionary process that slowly integrated the gastralia, the dorsal ribs, and part of the vertebrae and the pectoral girdle.

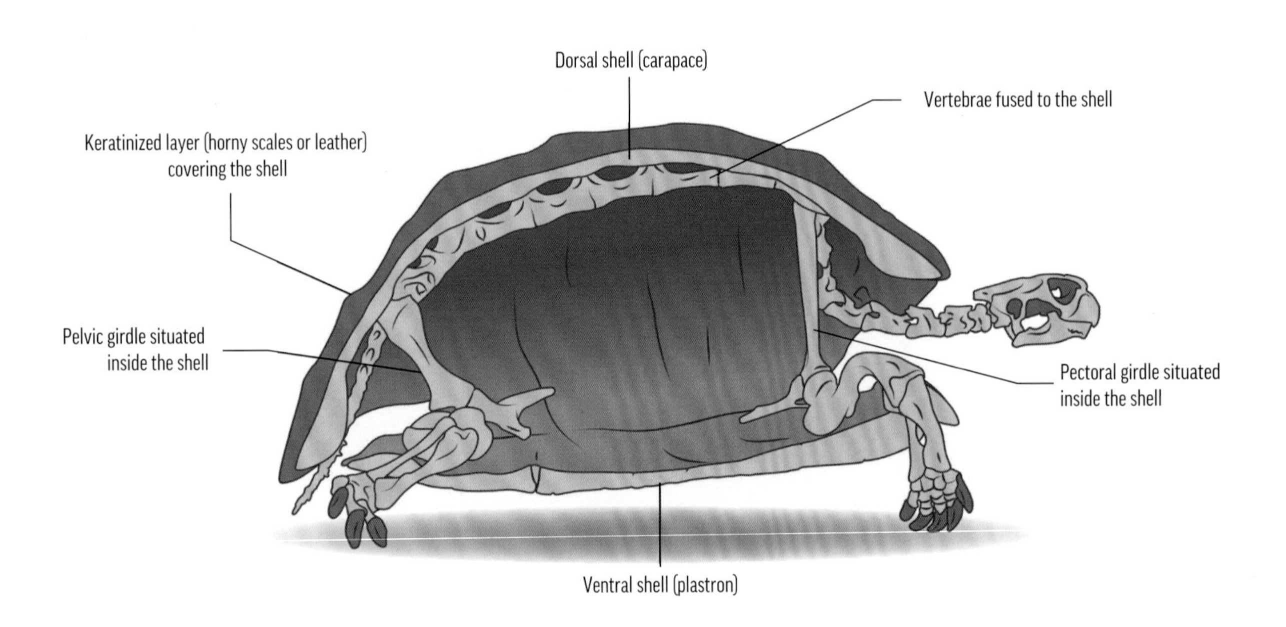

◄ Fig. 2.57. *Odontochelys*, an aquatic turtle from the Upper Jurassic in China, characterized by teeth and an incomplete shell, formed only by the plastron.

turtle was *Proganochelys*, from around 215 million years ago (the Upper Triassic) in Germany. *Proganochelys*, except for the vestigial teeth on the roof of its mouth, already possessed the whole range of "turtly" attributes, including a complete shell (both carapace and plastron; fig. 2.60). In 2008, however, the discovery of an even older turtle in China revolutionized our understanding of turtles' evolution. *Odontochelys*, as it has been named, had a quite astonishing construction: while it did have a solid plastron, it had no carapace, its ribs being large but not fused to one another (fig. 2.57). Moreover, as indicated by its name (meaning "toothed turtle"), its jaws were equipped with teeth! The icing on the cake, for paleontologists, is that its pectoral girdle was situated neither inside the rib cage, as in other turtles, nor around the rib cage, as in other classic tetrapods. Instead, it occupied a "middle" position: in front of the first ribs, which were enlarged (fig. 2.58). But the story does not stop there.

The discovery of *Pappochelys*, an even older turtle, from around 240 million years ago (the Middle Triassic) in Germany, once again contributed some key elements relevant to the exciting evolutionary history of the turtles' shells. *Pappochelys* had teeth in both jaws (like *Odontochelys*), had temporal fossae (we shall return to this), and completely lacked a shell. It did possess enlarged, and sometimes joined, gastralia (ventral ribs)—a prelude to the plastron—which served as

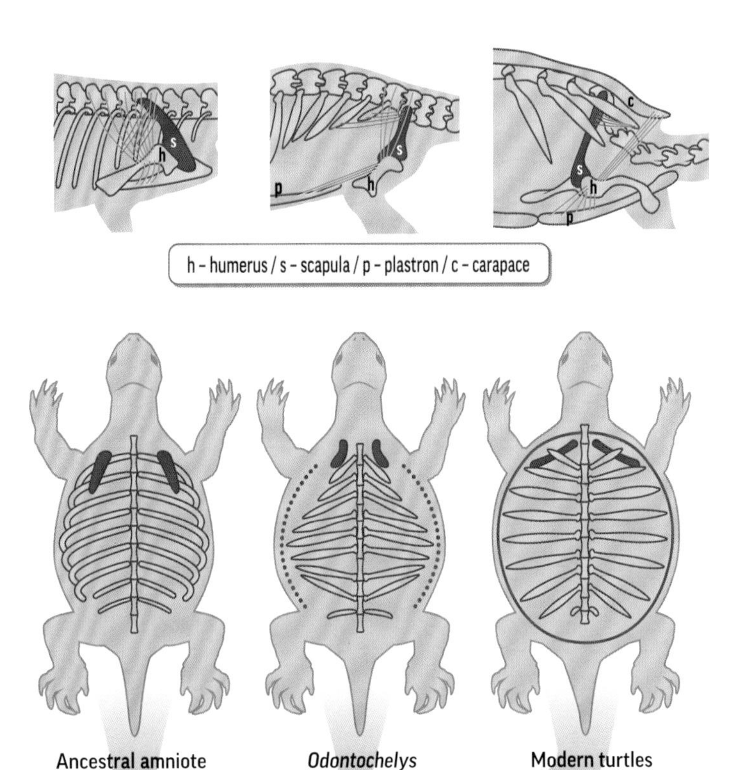

► Fig. 2.58. The appearance and evolution of the shell in turtles. *Odontochelys* (from the Triassic in China), with only a plastron, corresponds to an "intermediate" evolutionary stage between tetrapods without a shell and turtles with a complete shell (plastron and carapace).

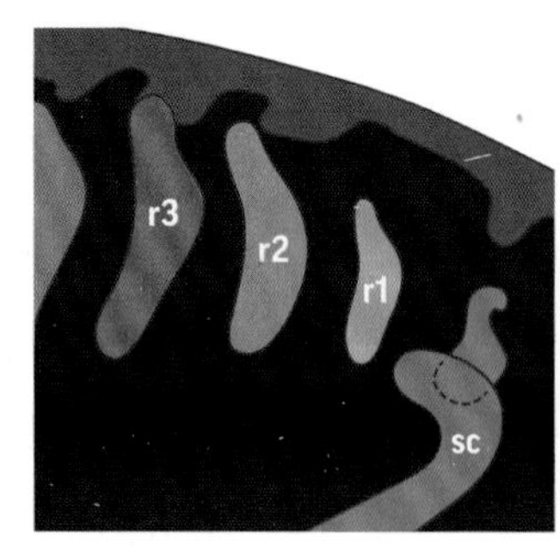

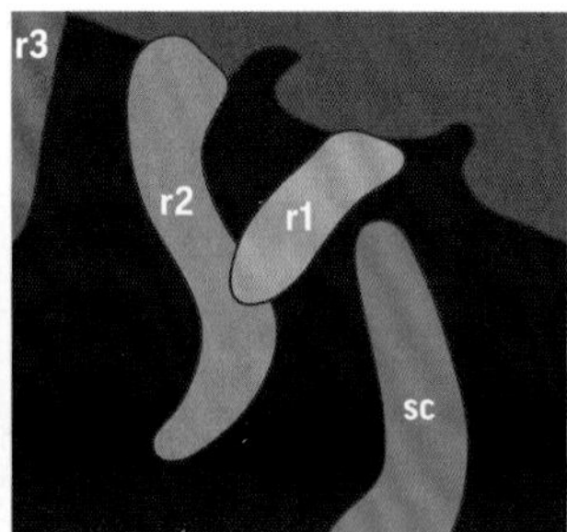

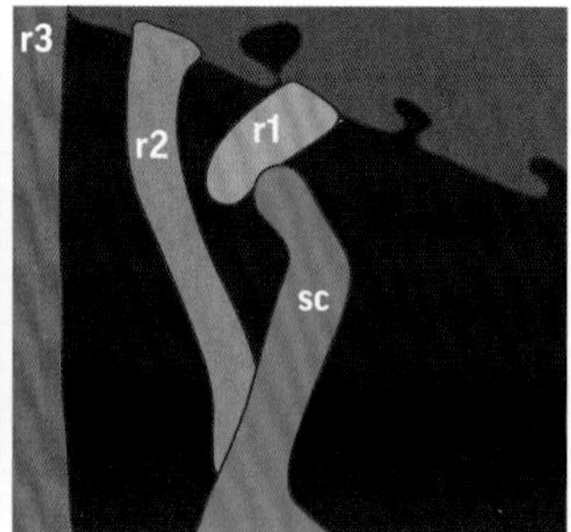

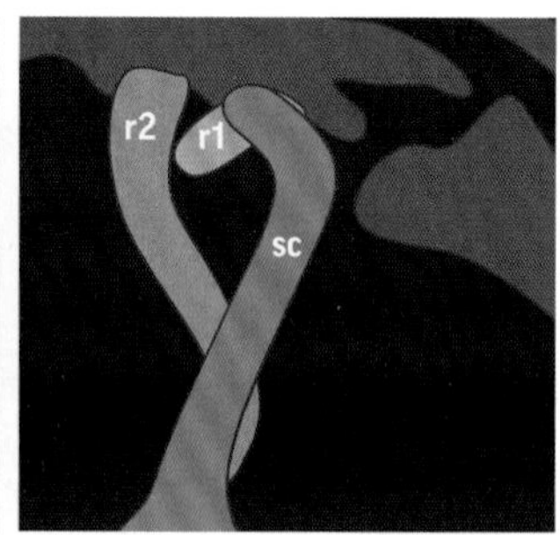

▲ Fig. 2.59. The migration of the scapula (bone of the pectoral girdle, labeled sc) from the outside to the inside of the rib cage during embryonic development of modern turtles.

▼ Fig. 2.60. The evolution of some key characteristics in cryptodiran turtles, starting with a form from the Permian and Triassic.

body armor (fig. 2.60). *Pappochelys*'s morphology therefore shows us that (a) the formation of the plastron in turtles is accomplished partly by fusion of the gastralia (just as the carapace is also formed partly by the fusion of the ribs) and (b) the formation of the plastron evolutionarily preceded the formation of the carapace.

While these paleontological discoveries were occurring, embryological studies were revealing that, in modern turtle embryos, during an early stage of development, the scapula (one of the bones of the pectoral girdle) migrates from a position that is anterior and external to the rib cage (similar to the position observed in *Odontochelys*) to an internal one, typical of turtles (fig. 2.59). This brilliantly illustrates the famous theory of recapitulation formulated by German biologist Ernst Haeckel (1834–1919), which stipulates that ontogenesis (an organism's development) recapitulates phylogenesis (the evolutionary history of which this organism is a product).

For most turtles, the outermost measure of defense is a layer of horny scales atop the shell (fig. 2.56). These scales are of varying thickness in different species.

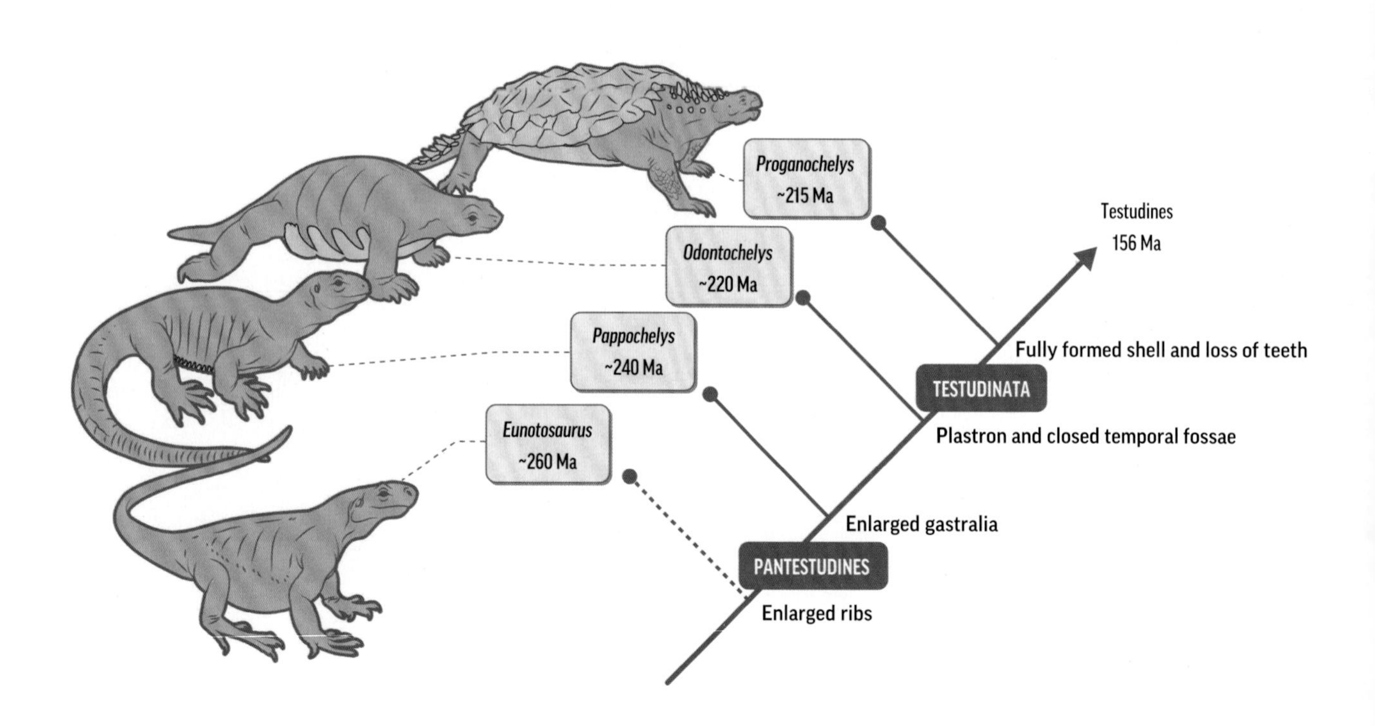

The Thorny Question of Origins

The origin of turtles and their kinship relations with other reptiles, as well as their ecological origin (marine vs. continental), are the subject of fierce debate.

After the discovery of *Proganochelys* (from ca. 215 Ma)—which, for a long time, remained the "grandmother" of all turtles—it was thought that turtles had evolved on land, given *Proganochelys*'s massive and heavy shell. But the successive discoveries of *Odontochelys* (from ca. 220 Ma) and *Pappochelys* (from ca. 240 Ma) upset this well-established notion by seeming to tip the scales toward an aquatic origin story. *Odontochelys* was found in marine deposits, but its skeleton does not exhibit any notable adaptations to the marine environment, so we may be looking at a terrestrial animal that was transported into a marine environment. As far as *Pappochelys* is concerned, it was found in a lake formation, where it was the most highly represented reptile. This seems to indicate that *Pappochelys* lived in the lake or on its banks. Perhaps, therefore, the plastron (resulting from the fusion of the thickened gastralia) developed not only as a protection but also as a ballast, to control buoyancy. Thickened ribs and gastralia have, in other cases, been shown to be in accordance with an aquatic or semiaquatic way of life and present in numerous such species. Considering these latest discoveries in combination, *Pappochelys* would seem to be a testament to turtles' aquatic origin and *Odontochelys* may represent their first incursion into a marine environment. Whatever the case may be, currently the turtles' most likely progenitor seems to be *Eunotosaurus*, from around 260 million years ago (the Permian) in South Africa. But *Eunotosaurus* was a land-dweller! So what happened between 260 and 240 million years ago? This is what paleontologists are trying to determine, each day unwinding this "Ariadne's thread" of turtles' evolutionary history a little more, with the aid of their discoveries (fig. 2.60).

Let us now confront the issue of turtles' origins. The classic reasoning is that, because they lack temporal fossae, turtles belong to the anapsids (see p. 26). In that case, the parareptiles of the Permian are the sister-group to the turtles, and the group that gave rise to the turtles is either the large, herbivorous pareiasaurids or the small procolophonids. On the other hand, studies founded on morphological data and those founded on molecular data seem to agree in considering turtles as having evolved from diapsids that lost their temporal fossae. The turtles' closest relatives among living diapsids, then, are the lepidosauromorphs (lizards and snakes) if one follows the morphological analyses and the archosauromorphs (crocodiles and birds) if one follows the molecular analyses. So far neither of these hypotheses has gained consensus, notably because of the absence of very ancient turtle fossils that would clearly attest to turtles' diapsid nature. But in this instance also, *Pappochelys* has managed to provide some clues. Its skull reveals two small temporal fossae, a closed upper one and an open lower one, thus confirming the hypothesis of turtles' diapsid origin. And the phylogenetic analyses undertaken during the examination of *Pappochelys* paint turtles as part of a sister-group to the sauropterygians, combined with whom they form a sister-group to the lepidosauromorphs.

The kinship relations among turtles are relatively simpler, at least at the higher levels of classification. If we exclude some primitive forms, turtles can be divided into two groups, characterized by the way they retract their neck into their shell: those of suborder Pleurodira retract it horizontally, sideways, into the shell, whereas those of Cryptodira fold their neck vertically while retracting their head backward (fig. 2.61). The overwhelming majority of today's turtles belong to the second group.

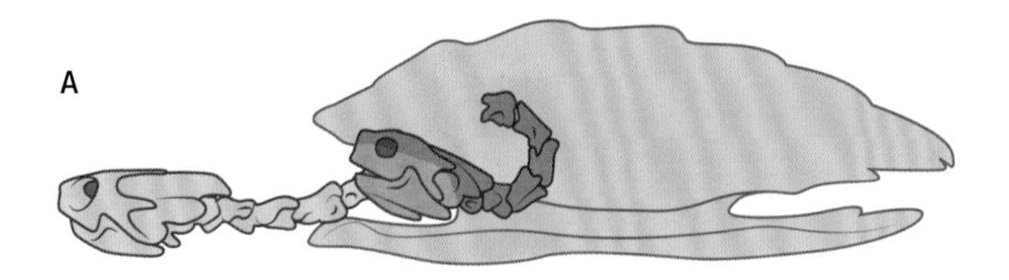

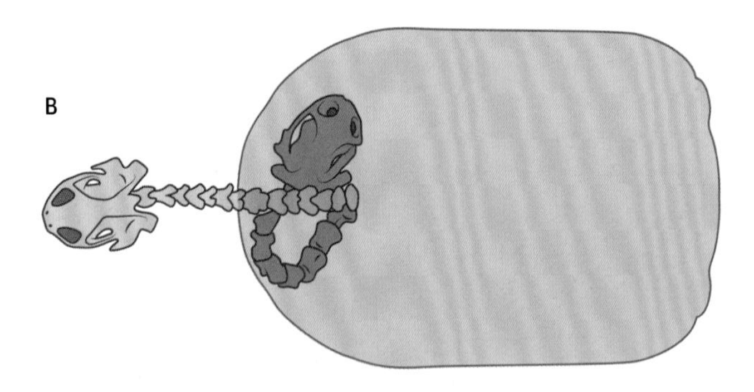

▲ Fig. 2.61. Types of neck retraction in turtles: vertical among cryptodirans (A), horizontal among pleurodirans (B).

▼ Fig. 2.62. A cladogram showing the principal marine invasions by cryptodiran turtles. Only marine turtles are represented.

The Armored Ones Get Wet

If we put the cases of *Pappochelys* and *Odontochelys* aside, the return to an aquatic and notably a marine environment occurred at least three times during turtles' evolutionary history: twice among the cryptodirans (during the Late Jurassic and the Early Cretaceous) and once among the pleurodirans (during the Late Cretaceous) (fig. 2.62). Some turtles remained coastal and amphibious, while others took to the open sea. The significant adaptations to the marine environment, such as swimming paddles, a salt gland, and the lightening of the shell (thanks to some large openings called fontanelles), have therefore appeared on multiple occasions, completely independently of one another. The high rate of evolutionary convergence among sea turtles shows the action of the environment on individual organisms in a spectacular fashion: given a certain environment with specific constraints, the field of possible adaptive solutions is limited; the pressure exerted by the environment then results in the selection of anatomical characteristics that are similar in each species, which in turn leads to the important resemblances we observe between groups that are phylogenetically fairly distant (see "Convergence," pp. 198–99).

Cryptodirans and Their First Forays into the Water

It was at the end of the Jurassic, between 155 and 145 million years ago, that three families of turtles (plesiochelyids, thalassemydids, and eurysternids) colonized the marine environment. Of small to medium size (20–60 centimeters), they had shells covered in scales (fig 2.63). Their limbs were fairly similar to those

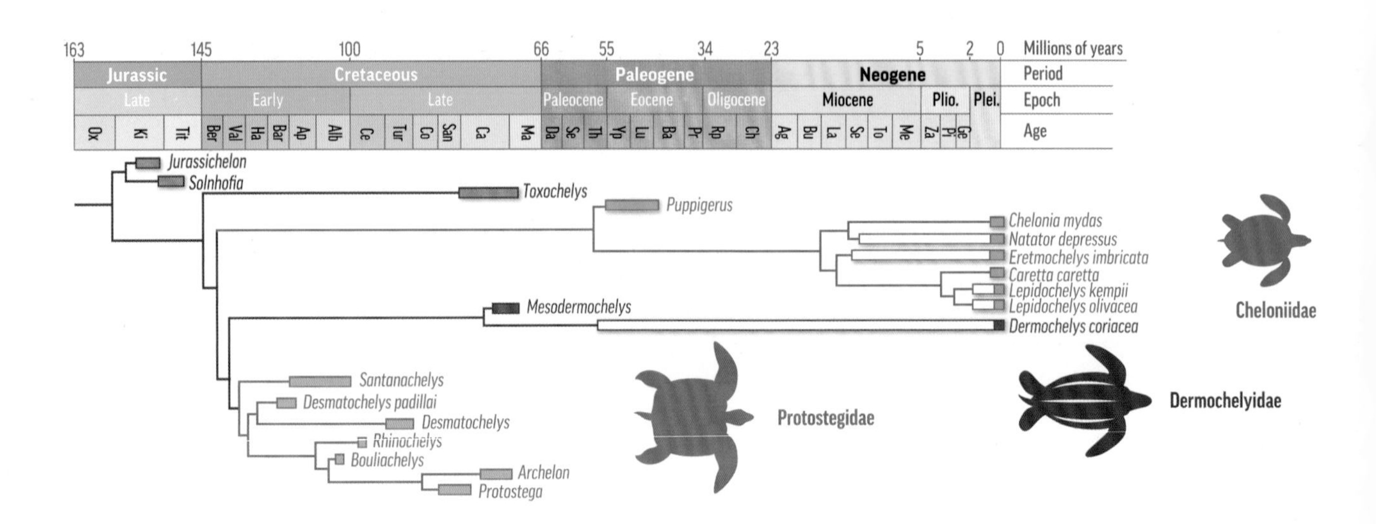

of today's freshwater turtles and didn't resemble swimming paddles at all: the forelimbs and hind limbs were large, equivalent in size, and equipped with articulated fingers and toes. Their swimming style must have involved paddling their four limbs alternately, as today's freshwater turtles do. They lived principally in coastal areas and, as shown by isotopic analyses, some occupied brackish environments or, alternatively, freshwater locales. They are known from only Europe and Argentina and disappeared at the very end of the Jurassic, when the coasts of the shallow epicontinental seas they inhabited shrank.

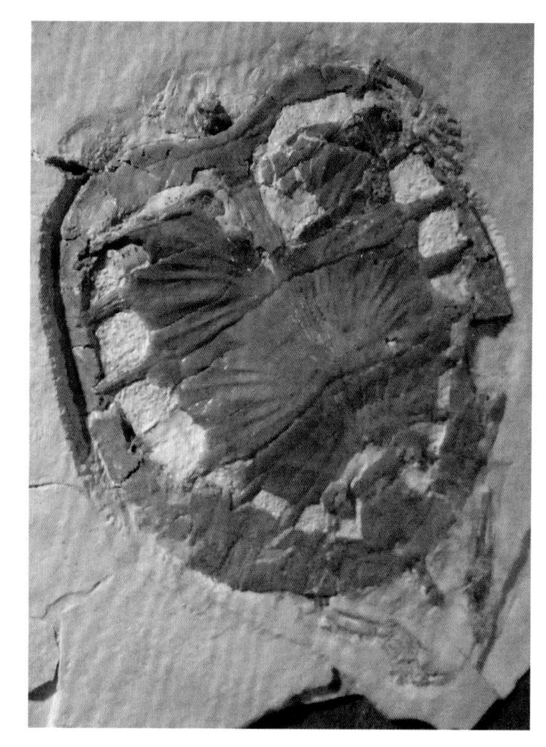

◄ Fig. 2.63. *Plesiochelys*, a nearshore sea turtle belonging to Cryptodira, from the Upper Jurassic in Europe.

Some Giants among the Turtles: The Chelonioids

It was once again among cryptodirans—during the Early Cretaceous, about 120 million years ago—that the chelonioids, or true sea turtles, first appeared. They rapidly diversified and spread throughout the seas, comprising two principal groups: the dermochelyoids (to which today's leatherback turtle belongs) and the chelonioids (which includes other modern sea turtles).

The chelonioids are one of the rare groups of marine reptiles, together with some crocodylomorphs, that survived the Cretaceous/Paleogene crisis. All of today's sea turtles belong to this large group, and although they are now represented by only two families, six genera, and seven species, sea turtles were of much wider variety in the past, especially the Cretaceous, their golden age (fig. 2.62).

Chelonioids were (and are) defined by their forelimbs: large, rigid, swimming paddles (fig. 2.64) used to propel them in underwater flight (see "Locomotion in Ichthyosaurs and Sauropterygians," pp. 148–49). Other morphological and physiological modifications allowed this group to evolve toward a pelagic lifestyle. The main adaptations were the reduction in the ossification of the shell thanks to the fontanelles (see "*Archelon*," p. 83, and *Allopleuron*, fig. 2.66), the lightening of the shell thanks to thinner horny scales than those of terrestrial turtles—or even replacement of those scales with leather (as in the leatherback turtle)—and the presence of an important opening called the interorbital foramen, to house a large salt gland.

Let us first deal with the dermochelyoids, and for a good reason: they are the oldest known chelonioids, documented since the Early Cretaceous (120–100 Ma) in Gondwana, comprising today's southern continents. The fossilized remains from that time belong to the family of the protostegids, which includes some standouts. *Desmatochelys padillai*, from the Barremian/Aptian boundary in Colombia, is considered the oldest "true" sea turtle. With a shell more than 1.5 meters long, it demonstrates that this family attained large sizes very early on! *D. padillai* was already equipped with large swimming paddles, and large openings were present in its shell, as in later protostegids.

At the other end of the spectrum of size, *Santanachelys*, from the Aptian/Albian boundary in Brazil, was only about 15 centimeters long. In its case as well, both carapace and plastron already had fontanelles but were covered in scales, while the forelimbs were short, with articulated fingers as in today's freshwater varieties. *Santanachelys*'s interorbital foramen, however, was already large and attests to the presence of a salt gland among the most ancient dermochelyoids. *Rhinochelys*, from the later Albian stage, had forelimbs that not only were longer than its hind limbs but, more importantly, ended in long, flat fingers, with conjoined phalanges making the whole structure more rigid—in other words, true swimming paddles. Its carapace was largely "windowed," and its plastron was light and star-shaped, typical of all dermochelyoids.

It was during the Late Cretaceous that, within this same family of the protostegids, there arose the gigantic *Protostega* and the more famous *Archelon*, the star of sea turtles, almost 3 meters long and 5 meters across (fig. 2.64). This giant, to which the leatherback turtle is closely related, lived 75 million years ago in North America's inland sea (the Western Interior Seaway), had very large, long front swimming paddles that allowed it to swim continuously while practicing underwater flight (like today's sea turtles), had a plastron that was typically star-shaped, and had a carapace that was considerably lightened by large fontanelles. The shell overall had no scales and probably was covered with leather, as is the leatherback turtle's. Its large, slender skull was more than half a meter long and ended in a strong, probably prehensile, beak, capable of shredding its prey. Because the inside of the modern leatherback turtle's mouth is lined with keratinized pointed structures resembling teeth, which help it hold on to its slithery prey—namely jellyfish—we might imagine that *Archelon* was similarly equipped. This colossal turtle probably actively fed on pelagic mollusks such as squid and perhaps, like the leatherback turtle, on jellyfish.

► 2.64. *Archelon*, a protostegid turtle from the Upper Cretaceous in North America. Peabody Museum of Natural History (New Haven, Connecticut, United States).

The protostegids declined at the end of the Cretaceous and went extinct during the Cretaceous/Paleogene crisis, 66 million years ago. From then on, the dermochelyoids were represented only by the family of the dermochelyids. While they are known starting from the Upper Cretaceous, around 85 million years ago, they reached peak diversity during the Paleogene period (66–35 Ma). These turtles are divided into two subgroups, with quite different shells: (a) those with a carapace lightened by large fontanelles and (b) those, such as *Dermochelys* (the leatherback turtle) and *Psephophorus*, from the Oligocene and Miocene epochs (30–5 Ma), with no keratinized scales on their carapace but with a skin stretched over a multitude of dermal bones that formed secondarily to constitute a very robust mosaic. The latter are somewhat strange, having first lightened their (internal) shell by means of apertures and the loss of keratinized scales, only to then reinforce it by covering it with a secondary (external) shell composed of osteoderms!

During the Late Cretaceous there were some other really strange giant turtles, known from phosphate deposits in Morocco (see chapter 6, p. 172). *Ocepechelon*, a dermochelyoid, was a distant cousin of the leatherback turtle. Only a single skull has been discovered, but it is a massive 70 centimeters long. On that basis, the animal's estimated overall length of 3.5 meters would make it one of the largest turtles ever, longer than *Archelon*. Uniquely among turtles, *Ocepechelon*'s nostrils were so high up as to be almost between the eyes (like a whale's blowhole), a characteristic typical of pelagic (open-ocean) animals (fig. 2.65). What is most remarkable, however, is its very long bony snout, shaped like a pipette. Not only no other turtle—living or extinct—but no other tetrapod has such a snout. Reminiscent of a seahorse's, it has been interpreted as meaning that this turtle ate by sucking up its food. *Ocepechelon* was therefore probably

Archelon

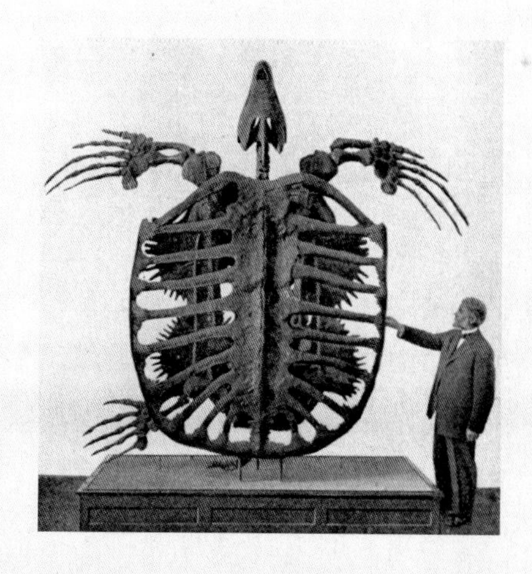

Archelon, from the Campanian stage (80 Ma) in the United States, is the undisputed star among sea turtles. Its name means "ancient turtle." *A. ischylos* is the only species of this giant turtle, discovered in South Dakota at the end of the nineteenth century. One of the first specimens found (shown here, in the Peabody Museum at Yale University) is spectacularly complete, except its right rear swimming paddle is missing. This has led people to wonder: Was the right hind limb taken by a scavenger after the animal's death? Or was it bitten off in life by a large predator, such as shark or a mosasaur? The mystery remains. Another fantastic *Archelon* specimen is displayed at the Natural History Museum in Vienna (Austria).

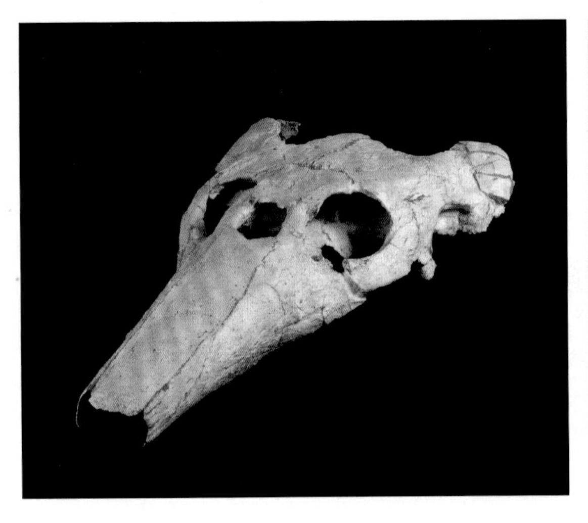

▲ Fig. 2.65. *Ocepechelon*, a giant dermochelyoid turtle with a 70-centimeter-long skull, from the end of the Cretaceous in Morocco. Skull, OCP Group (Khouribga, Morocco).

nothing more than a peaceful giant, feeding on small invertebrates and fish.

Let us move on to the second group of Chelonioidea: this group, specifically the family Cheloniidae, comprises fossil forms and all modern sea turtles except the leatherback.

Cheloniids probably arose during the Albian age (110 Ma), but it was during the Late Cretaceous and the Paleogene that they diversified and spread widely. Cheloniid fossils from this period have come to light basically everywhere in the world. Nonetheless, if we look a little closer (at the level of genus and species), the classification schema changes considerably. Unlike their current representatives, fossil species of Cheloniidae seem to have been somewhat restricted in their geographic range, because it is unusual to find the same species in different deposits. They came in many shapes and sizes, from less than half a meter to more than two and a half meters, as in the case of *Allopleuron*, from the Maastrichtian stage in the Netherlands and Belgium. *Allopleuron* (fig. 2.66) was equipped with a very low-profile and hydrodynamic shell as well as large swimming paddles, which tells us it probably lived in the open ocean. Other cheloniids, such as *Euclastes*, one of the rare taxa known from both below and above the Cretaceous/Paleogene boundary, were of average size and mostly stayed close to shore, as attested by their less hydrodynamic shells and their shorter and less rigid paddles. This genus is also characterized by a low and short skull and very extensive crushing surfaces on both the roof of the mouth and the front of the mandible, features that point to a diet of tough, hard foods.

► Fig. 2.66. *Allopleuron*, a cheloniid turtle from the end of the Cretaceous in the Netherlands and Belgium, Natuurhistorisch Museum Maastricht (the Netherlands). It was found in the same stratum as the first mosasaur described by Georges Cuvier.

The Unique Marine Incursion of the Pleurodira

The bothremydids, the only seagoing family of pleurodirans, first appeared around

130 million years ago (the Barremian age of the Early Cretaceous) in Gondwana (they are known from Brazil). Equipped with an oval shell devoid of fontanelles and covered in scales, some seem to have had a salt gland, although one that was situated differently than in cryptodirans. Bothremydids' forelimbs and hind limbs resembled those of freshwater turtles.

The bothremydids were very widespread during the Late Cretaceous and especially the early to mid-Paleogene, or roughly 80 to 40 million years ago. The specimens found in the phosphates of Morocco (see chapter 6, p. 172) attest to a wide range of species, which included (as did Chelonioidea) turtles that were "crushers," with compact skulls and sharp jaws, and those with a very long and slender snout, such as *Labrostochelys* (fig. 2.67), which must have used that snout to suck up food. Like some chelonioids and some dyrosaurid crocodiles, the bothremydids did not seem to suffer at all during the Cretaceous/Paleogene crisis. Conversely, they took advantage of it and further diversified. They likely died out around 40 million years ago, in the mid- to late Eocene.

▲ Fig. 2.67. *Labrostochelys*, a bothremydid turtle from the Paleocene in Morocco.

The Squamates: Sea Serpents and Leviathans

The squamates (order Squamata), which include mosasaurs, snakes, lizards, and the unusual amphisbaenians, are peculiar diapsid reptiles in that their lower temporal fossa is no longer enclosed, owing to the loss of the quadratojugal bone (in fig. 2.2, the triangular bone below and to the rear of that opening). They are characterized especially by the fact that their quadrate bone (in fig. 2.2., the small bone at the rear of the cranium, next to the quadratojugal bone), which forms an articulation with the mandible, is mobile, a feature called **streptostyly**. Streptostyly allows for significant movements of protraction and retraction of the lower jaw relative to the cranium, thus enabling the ingestion of large prey. In addition, most of the bones of squamates' skull are articulated and slide on one another, rather than being fused. Called **kinesis** (fig. 2.69), this mobility allows the skull to expand, helping the animal literally get its head around big mouthfuls. This adaptation is pushed to the extreme by macrostomatan snakes (e.g., pythons and boas), which use it to swallow prey much bigger than their skull, such as deer, goats, and pigs. In mosasaurs, while the streptostyly of the quadrate was still present, in larger animals, such as *Mosasaurus* (figs. 2.84 and 6.13) and *Prognathodon* (fig. 6.11, pp. 176–77), the kinesis was nonexistent, showing that these mega-predators no longer that ability because they were sufficiently bigger than their prey. Last, squamates are often

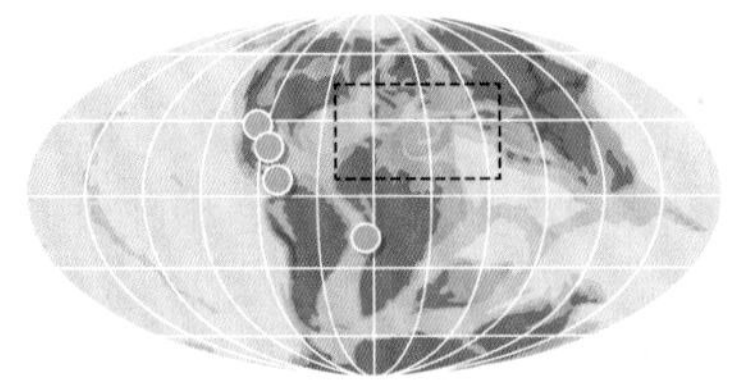
Beginning of the Late Cretaceous

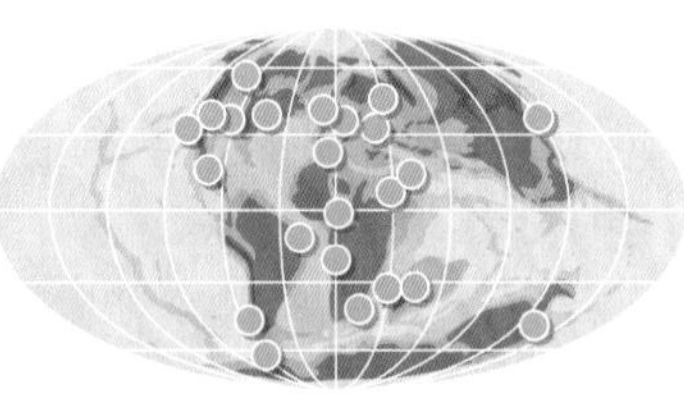
End of the Late Cretaceous

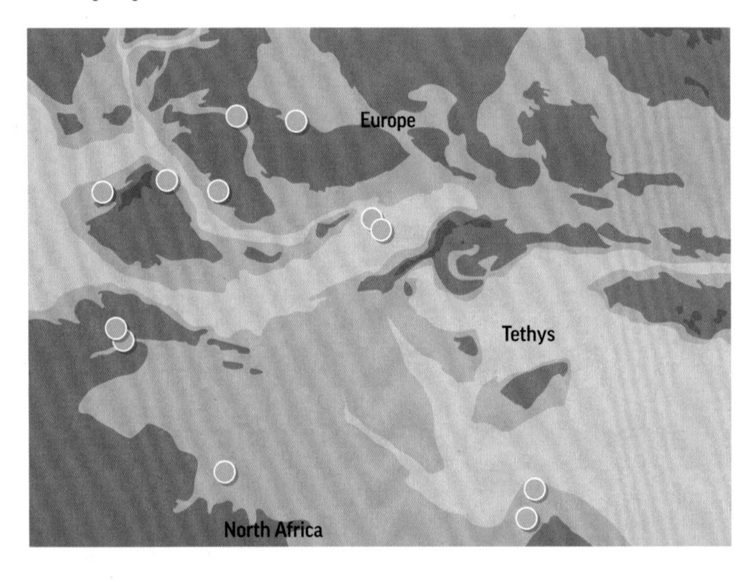

▲ Fig. 2.68. The distribution of marine squamates during the Cretaceous, with a focus on the Mediterranean Tethys.

equipped with teeth that curve rearward, like hooks, on their palates (on the pterygoid bone; fig. 2.79), which act like harpoon points, preventing struggling prey from moving away from the throat: a redoubtable trap with no escape!

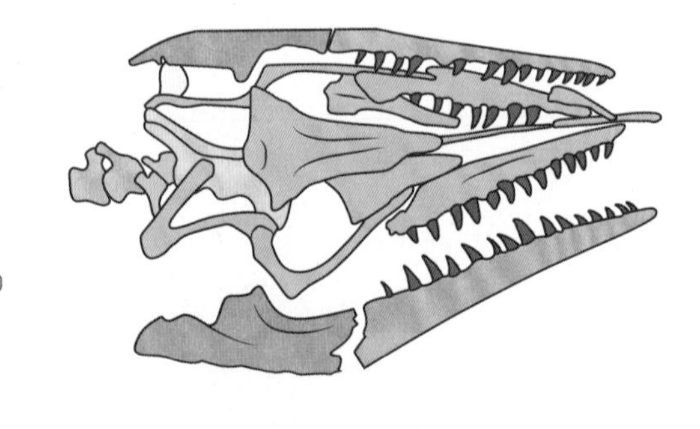

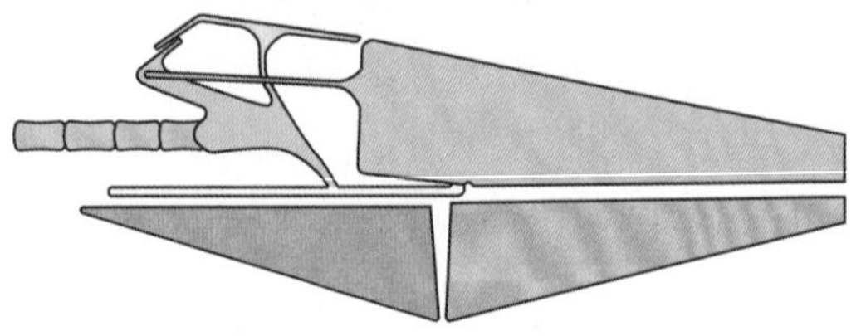

► Fig. 2.69. **Below**, the divisions of the skull and mandible capable of movement relative to one another, allowing for increased flexibility in mosasaurs. **Above:** Since the bones were not fused to one another, the skulls of squamates are often preserved broken up, as in the case of this mosasaur.

Squamata is a very ancient order (squamates first appeared about 170 Ma), and today, with nearly eleven thousand species represented, it is the second most diverse group of vertebrates, after the **teleost** fish (essentially those belonging to Osteichthyes, the bony fish). It is also the group in which living reptiles vary most in size, from less than 20 millimeters to more than 9 meters in length! The great majority of today's squamates are terrestrial, and the occasional marine forms are limited to several families of snakes and to the marine iguana of the Galapagos (see chapter 1, p. 8). However, this was not always the case. From the beginning of the Late Cretaceous and over a relatively "short" time (100–66 Ma), squamates underwent a spectacular proliferation (paleontologists speak of **evolutionary radiation**)—both in the number of species and in the variety of ecological niches they occupied—in the seas. Over less than 10 million years, two major groups, with very different destinies, diversified: the ophidiomorphs and the mosasauroids. This very important branching is, to say the least, surprising, because, while the mosasaurs continued their diversification and maintained a worldwide presence for the rest of the Late Cretaceous, the ophidiomorphs' foothold in the marine environment did not last for long or spread very far (essentially it lasted from the Cenomanian age to the Turonian age and was limited to the Mediterranean Tethys; fig. 2.68). In addition, this radiation would be the last that the Mesozoic marine reptiles would experience. This is how the squamates became the first group of marine reptiles to be given a historical description (because of a mosasaur) and the last to have invaded the oceans!

The Complicated Relations among the Squamates

The kinship relations among today's squamates remain blurred, particularly as concerns the position of the snakes, even when

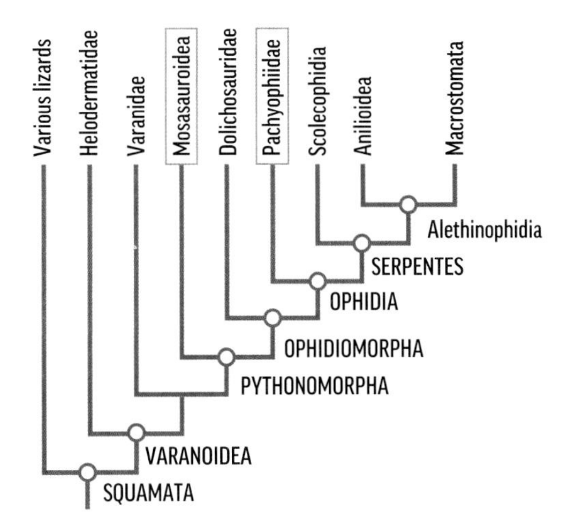

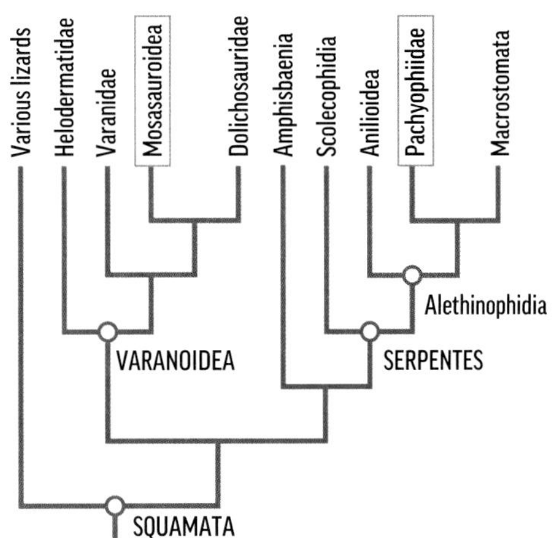

◄ Fig. 2.70. Simplified phylogenetic positions of the mosasauroids, considered as close both to varanoid lizards (right) and to snakes, with which they form the group Pythonomorpha (left), and of the snakes with legs (pachyophiids), considered both as basal forms (on the left) and as evolved forms of macrostomates (on the right).

fossils are part of the phylogenetic analyses. While even the inclusion of snakes among the "ophidiomorphs" (see below) will no longer cause much of a stir, the position of the ophidiomorphs and the mosasauroids among the squamates has been debated for more than a century. The mosasauroids (including the "aigialosaurids" and mosasaurids) are considered by some as being closer to lizards (like the monitor lizards)—thus once more giving credence to French anatomist Georges Cuvier's (1769–1832) original vision (see "Georges Cuvier," p. 95)—and by others as closer to snakes, being grouped with them within the pythonomorphs, as nineteenth-century American paleontologist Edward Drinker Cope held they should be (fig. 2.70). The "snakes with legs" (pachyophiids) are, for their part, considered both to be basal snakes to some degree and to have evolved from macrostomatan snakes (pythons, boas, etc.) (fig. 2.70). To this day, paleontologists debate these questions with each other peacefully, using scientific articles as their weapons.

These different outlooks are not without importance, to the extent that they bear on the issue of snakes' origins and why they acquired their distinctive body shape. Were the first snakes burrowers (i.e., terrestrial), or were they swimmers (i.e., aquatic)? Whichever is the case, snakes subsequently conquered a range of ecological niches both in oceans and on continents. This is how several lineages adapted independently and secondarily to the marine environment. Today's sea serpents belong to various groups of elapids (see chapter 1, p. 8). Between the Cenomanian (roughly 95 Ma) and the end of the Paleogene (roughly 34 Ma), snakes that adapted to the aquatic environment include the nigerophiids of Africa and Asia; the russellophiids of the Sudan, France, and India; and the paleophiids, some of which, such as *Palaeophis colossaeus*, from the Lutetian stage of the Eocene series in Mali, exceeded 9 meters in length.

Some Misunderstood Basal Forms among the Ophidiomorphs

The status of ophidiomorphs as a group remains the subject of discussion, and it includes snakes and some basal forms that are believed to belong to the lineage of

▲ Fig. 2.71. *Coniasaurus*, an ophidiomorph squamate from the Upper Cretaceous in the Mediterranean Tethys, as well as North America's inland sea (the Western Interior Seaway).

▼ Fig. 2.72. *Pachyrhachis*, a snake with legs (hind limbs!) from the Upper Cretaceous in the Middle East.

snakes; the latter, in the past, were grouped under the heading "dolichosaurs." These were small animals, less than a meter long, with a very long body and a long, laterally compressed tail, as well as short limbs (especially the forelimbs). They diversified significantly between 100 and 95 million years ago at middle latitudes. They would subsequently become rare, and their last representatives are known from the Western Interior Seaway (North America's inland sea) around 85 million years ago. They inhabited mostly the shallow coastal environments of the Mediterranean Tethys and are generally distinguished by an increase in their bone mass (see "The Secrets of Bone," p. 129). They are understood to have been not very active swimmers that would dive in shallow waters and therefore lived in circumstances relatively unfavorable for their wider distribution. Contrary to these tendencies, *Coniasaurus* (fig. 2.71) and *Dolichosaurus*, which do not exhibit this bone specialization, were found both in the Mediterranean Tethys and the Western Interior Seaway. They were probably surface swimmers that lived in a more open environment, which aided their capacity to disperse.

Eupodophis

Eupodophis, which means "snake with incomplete legs," is known from the Cenomanian stage (ca. 95 Ma), in Lebanon. Because of its macrostomatan skull (capable of extreme opening of the jaws), it may have evolved from the ophidians. The hind limbs are an ancestral trait in this group of squamates. The specimen shown is one of only three specimens of "snakes with legs" (the others belong to *Pachyrhachis* and *Haasiophis*) in which the limbs have been preserved.

The specimen, which is preserved across two slabs of limestone, is almost complete: a part of the spine (an estimated forty vertebrae) is missing, and the rear part of the animal is preserved next to the skull. This individual is estimated to have been about 85 centimeters long. The tail was very short (about 5 centimeters) and had been transformed into a swimming paddle. The hind limbs were shrunken (about 2 centimeters) and anatomically incomplete, with a reduced number of bones in the ankles and an absence of phalanges.

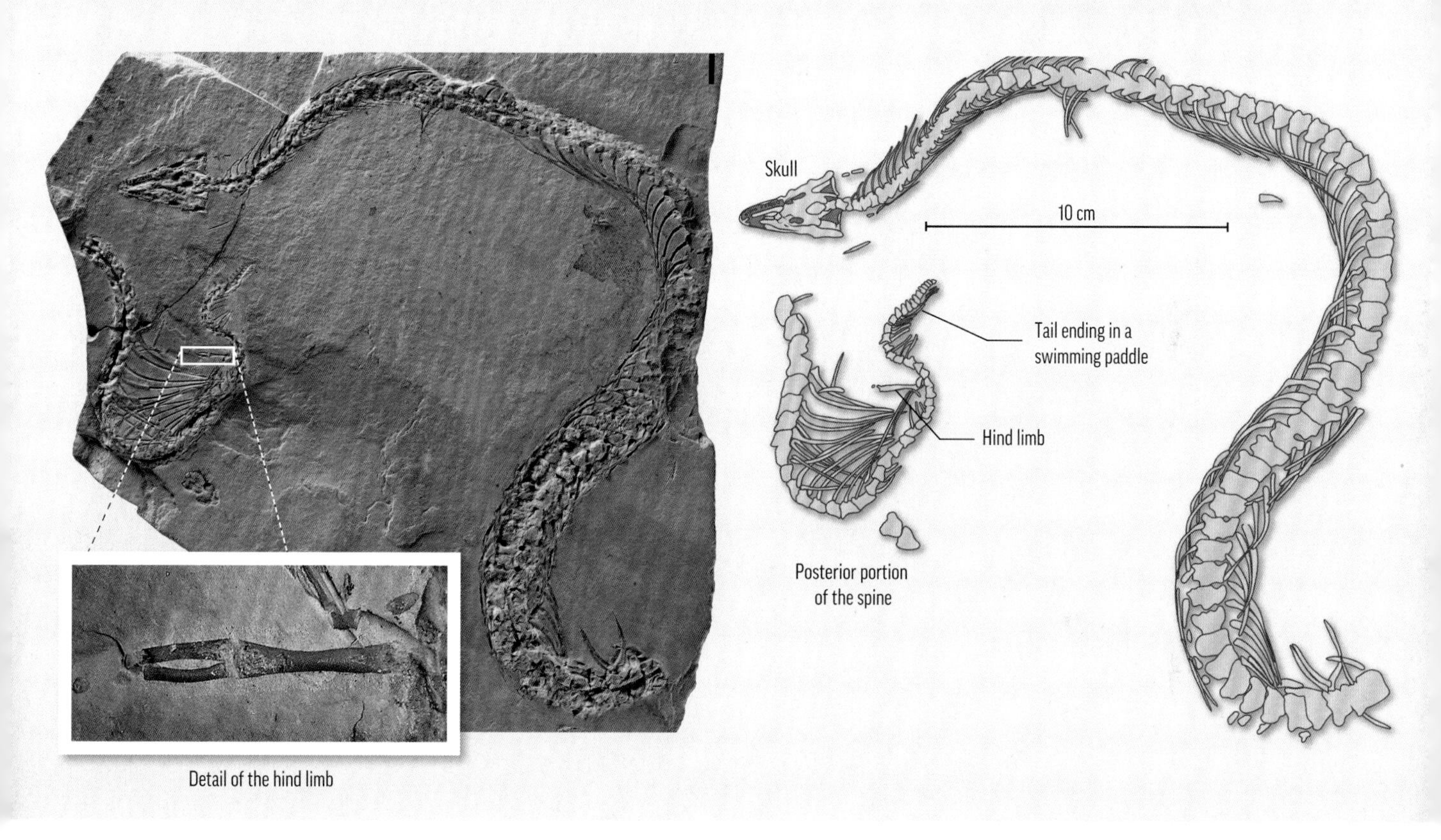

Did You Say "Snakes with Legs"?

The pachyophiids, or "snakes with legs," known from six genera, such as *Pachyrhachis* (fig. 2.72), inhabited only the tropical and shallow waters of the Mediterranean Tethys during the Cenomanian (about 95 Ma). They could reach 1.5 meters in length, were characterized by a long body and a short tail that was very flattened laterally (to serve as a swimming paddle), and possessed small but unmistakable hind limbs (figs. 2.72 and 2.73). Among today's snakes, vestigial remnants of the pelvic girdle and hind limbs can sometimes be observed in the skeleton (e.g., in pythons), but no form shows well-developed limbs. All pachyophiids had a pachyosteosclerotic bone structure, for increased

▲ Fig. 2.73. *Eupodophis*, a snake with legs from the Upper Cretaceous in Lebanon. Private collection.

weight (see "The Secrets of Bone," p. 129), and they were probably slow-swimming ambush hunters. *Tetrapodophis*, which was equipped with short forelimbs and hind limbs, was terrestrial and is not part of the pachyophiids.

From Coast-Dwellers to Leviathans: The Mosasauroids

What are mosasauroids (or mosasaurs)? On this subject, which has been debated for a very long time, today paleontologists agree to consider mosasauroids as a large group that includes the species classically labeled aigialosaurids and mosasaurids. Whether they are closer to lizards (particularly monitor lizards) or to snakes, on the other hand, remains a topic of debate (fig. 2.70), even though proponents of the first hypothesis seem to have regained some ground in recent years.

During the entire Late Cretaceous—in other words, for 30 million years—mosasauroids diversified rapidly. More than thirty genera and seventy species are known, which makes them the group of marine squamates that flourished the most, in terms of both number of species and ecological variety. They disappeared at the end of the Mesozoic, during the Cretaceous/Paleogene crisis, 66 million years ago. The fact that they were still well diversified and broadly distributed at the end of the Cretaceous points to the fact that they were not on a path to extinction. They were probably affected by the food-chain disruption—which, as mega-predators (at the apex of their food chains), they were particularly exposed to—caused by the catastrophes that characterized this crisis.

Mosasauroids had an essentially "monitor-like" body: very elongated, with a large skull and a long tail, usually compressed laterally; overall they present a very hydrodynamic aspect (fig. 2.75). How they reproduced remained a mystery for a long time, but discoveries in the first decade of the twenty-first century show that, from the very beginning, they were viviparous, and therefore they were not tied to a terrestrial environment. Mosasauroids came in three body types: (a) plesiopedal and plesiopelvic, (b) plesiopedal and hydropelvic, and (c) hydropedal and hydropelvic, with both microanatomical (fig. 2.76)

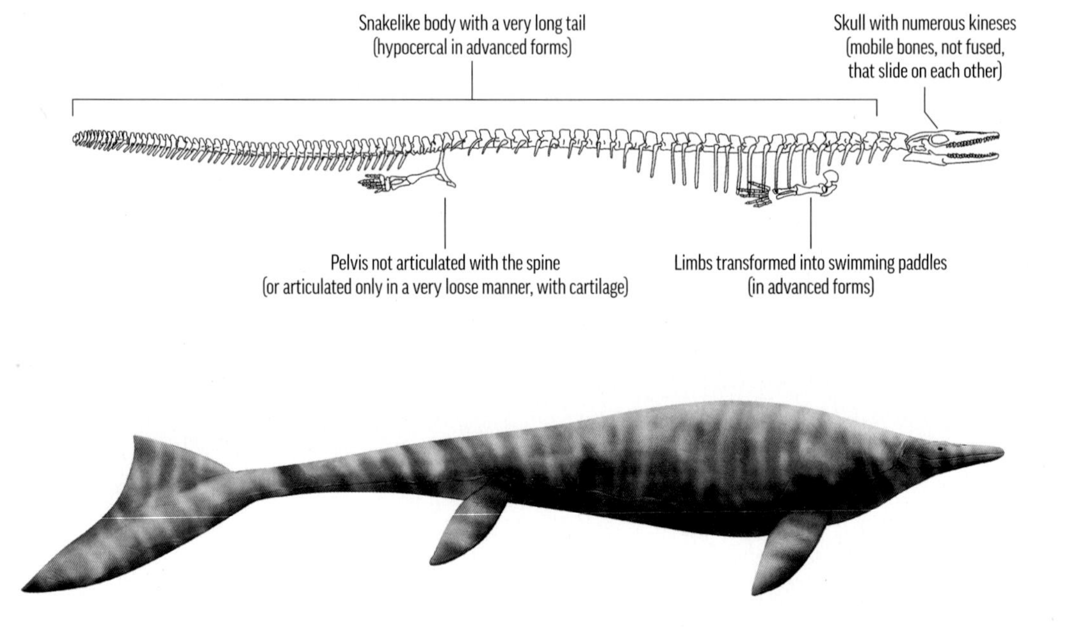

► Fig. 2.74. Skeleton of *Dallasaurus*, a primitive mosasaur from the Turonian in Texas (United States), showing its principal anatomical characteristics, some of which were modified in later mosasaurs (fig. 2.75).

► Fig. 2.75. *Plotosaurus*, a late mosasaur from the Maastrichtian in California (United States).

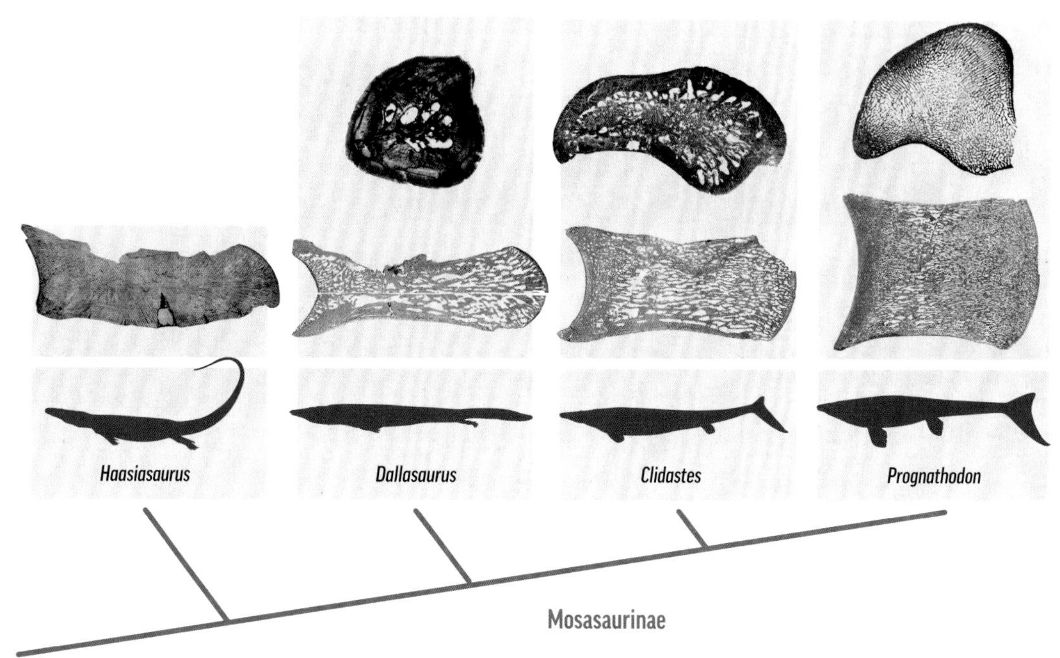

◄ Fig. 2.76. A progressive ecological adaptation to active swimming in the open ocean, shown by the internal structure of mosasauroid bones (cross section of humerus above, longitudinal cross section of vertebra below). The bony structures transitioned from osteosclerotic to spongy (see "The Secrets of Bone," p. 129).

and morphological (fig. 2.77) features that testify to an ever more advanced adaptation to a pelagic way of life. The plesiopedal and plesiopelvic mosasauroids retained "terrestrial" limbs (i.e., of the walking type) and a sacrum, a section of fused vertebrae connecting the pelvis to the spine. We are talking about small, not very active swimmers, 1 to 2 meters in length, that frequented shallow waters on the northern and southern edges of the Mediterranean Tethys around 95 million years ago. The plesiopedal and hydropelvic mosasauroids (which had limbs but had lost their sacrum) were slightly larger (1.5–3 meters long) (fig. 2.74). Known from around 90 million years ago, they were more widely distributed, since they have been found in Europe, Africa, and North and South America. These were more active swimmers, capable of dispersing over long distances. Last, the hydropedal and hydropelvic mosasauroids, the "true" mosasaurs, were large animals (3–15 meters long) equipped with large steering swimming paddles, each characterized by five fingers with phalanges that either were very long or numbered more than usual (cf. "From Legs to Swimming Paddles," p. 39).

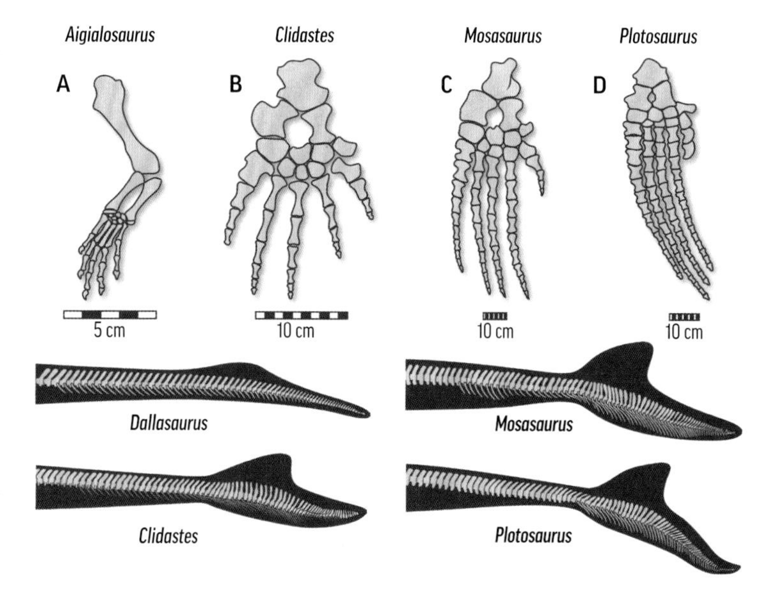

► Fig. 2.77. A diversity of front swimming paddles and caudal fins in mosasauroids, exhibiting the entire spectrum of ecological adaptations, going from small coastal forms (*Aigialosaurus*, *Dallasaurus*) to the large cruisers of the high seas (*Mosasaurus*, *Plotosaurus*).

Tethysaurus

▲► Fig. 2.78. *Tethysaurus*, a mosasaur from the Upper Cretaceous in Morocco. Skull, Muséum national d'Histoire naturelle (Paris, France).

Tethysaurus ("lizard of the Tethys") was a mosasauroid, known from the Turonian stage (ca. 90 Ma) in Morocco. *Tethysaurus* therefore belongs to the first known mosasaurs. While still equipped with terrestrial limbs, it probably did not have a sacrum; since it was about 3 meters long, it represents the perfect link between the first forms, which were of modest size (1–2 meters), with both "terrestrial" limbs and a sacrum, and the large, highly evolved mosasaurs (up to 15 meters in length), with swimming paddles and no sacrum.

Tethysaurus's skull, about 30 centimeters long, was slender, and its jaws were filled with many small, narrow teeth, which curved toward the rear, like harpoons. It probably fed on small invertebrates and fish.

The Turonian rock around Guelmim, in Morocco, is characterized by calcareous (chalky) concretions rich in often exceptionally preserved fossils of both vertebrates and invertebrates, something that allowed for two important discoveries concerning the mosasaurs. First, a very small specimen was found associated with an adult *Tethysaurus* in the same concretion. The anatomical, microanatomical, and histological characteristics all suggest that this is a juvenile specimen of *Tethysaurus* and not some prey. It represents one of only a few juvenile mosasaurs ever found. Second, this exceptionally conserved specimen allowed for the detailed description of one of the few known endocraniums of mosasaurs. In the controversy regarding mosasaurs' kinship relations within Squamata (i.e., are they closer to lizards or snakes?), it is significant that *Tethysaurus*'s endocranium resembles that of today's monitor lizards (see p. 87).

The shape of mosasauroids' tail—and therefore their swimming style—also changed over time: from long and laterally compressed in the most primitive forms, their tail became hypocercal, almost like that of ichthyosaurs, thus facilitating a transition from an anguilliform swimming style to a more efficient, quasi-thunniform one (see "Adaptations to the Aquatic Environment," pp. 118–19). They are regarded as active open-ocean swimmers capable of traversing wide expanses, as attested by their fossils, which have been found on every continent (including Antarctica), signifying their global distribution at the end of the Cretaceous.

Mosasauroids' skull provides important clues about their diet. First, something we have already underscored: the combination of cranial and mandibular kineses (fig. 2.69), powerful jaws with curved teeth, and a palate also equipped with teeth (fig. 2.79) allowed these predators to grip and then ingest very voluminous prey very efficiently, similar to snakes. Among the larger varieties of mosasauroids, however, these kineses had a tendency to disappear; these animals were so large that they could shred and swallow basically any kind of prey without a need to stretch! Second, mosasaurs are probably the reptiles with the teeth that are most revealing: often diagnostically very useful, allowing for the identification of a specimen down to the species level, they come in a very considerable range, which provides for a mostly open window onto the group's ecology. Mosasaurs could pierce, cut, or crush their prey (fig. 2.80). Small, narrow, pointed, and very overlapping teeth acted like a trap and, when associated with a long and slender skull, point to a diet of small

prey with soft bodies (fish, invertebrates with no shell, etc.), as is the case with *Halisaurus* and *Plioplatecarpus*. A large, long, and massive skull, armed with knifelike serrated teeth, is the hallmark of opportunistic mega-predators that must have shredded their prey, in the manner of *Mosasaurus* or *Tylosaurus*. Finally, a short and robust skull with large, short, bulbous teeth, as in *Globidens*, points to the use of those teeth to crush and grind shelled invertebrates (ammonites, bivalves, etc.). There are certainly many "intermediate" forms, such as *Prognathodon* (fig. 6.11, p. 176–77), a mega-predator with

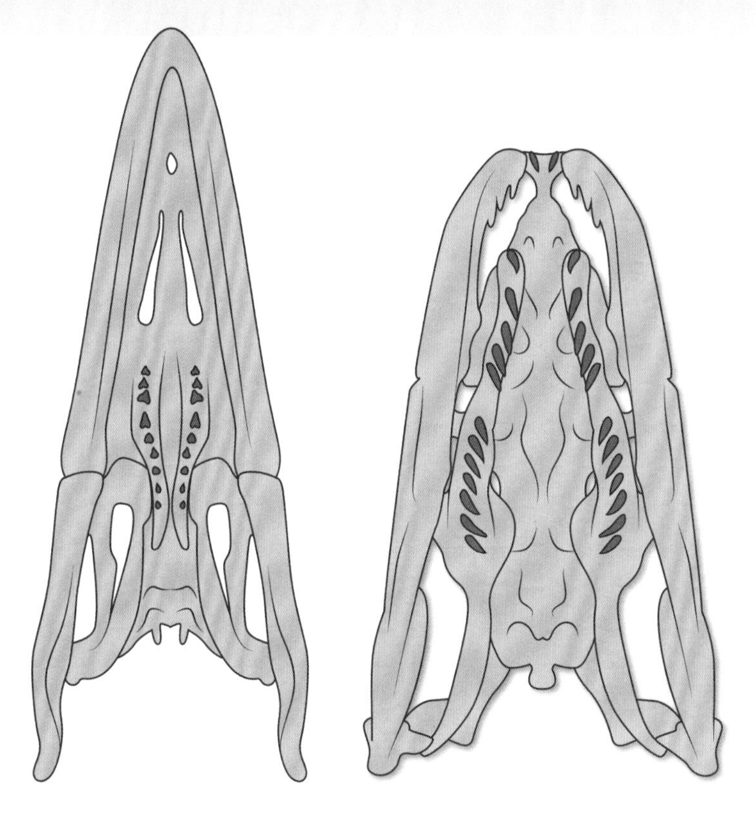

► Fig. 2.79. Details of the palate of a mosasaur (left) and of a snake (right), showing pterygoid teeth for efficient ingestion of prey.

a strong skull and large conical teeth with blunt ends, which must have fed on turtles and other large armored creatures. Or consider *Carinodens* (fig. 6.11), a small, slender mosasaur with short teeth, flattened laterally, which must have fed on invertebrates with fairly tough shells, such as crustaceans. We could provide many other examples. Everything was possible within this group of predators!

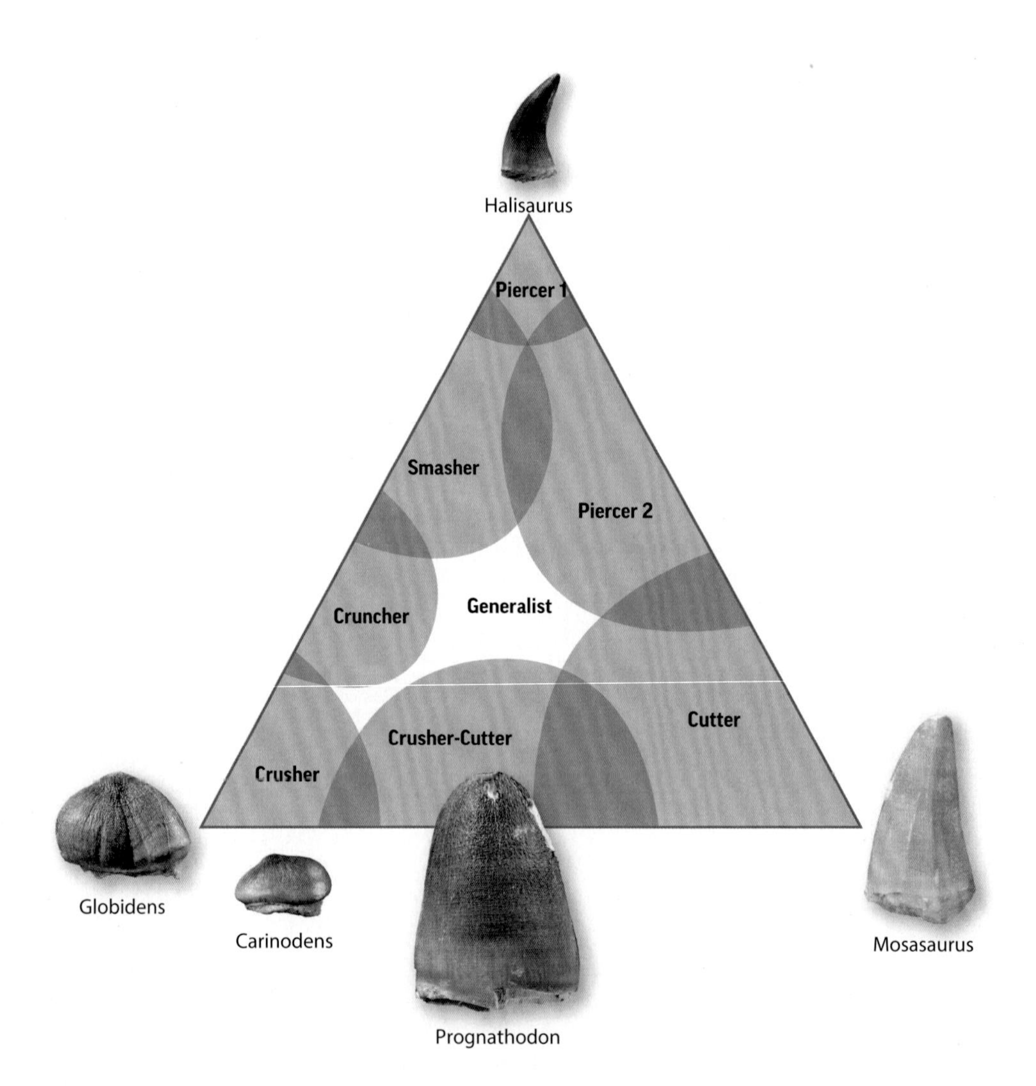

▶ Fig. 2.80. The study of the shape and size of mosasaurs' teeth provides insight into their probable diets: from the shape, we can deduce the function.

Georges Cuvier, "The Antiquarian of a New Age"

▲ Fig. 2.81. Georges Cuvier, by François-André Vincent.

During the conquest of Maastricht (in the Netherlands) in November 1794, the French revolutionary army requisitioned—among other war trophies, both artistic and scientific—some gigantic jaws almost 1.3 meters long, belonging to a creature referred to in naturalist circles as the "Maastricht unknown." A very vague name, but one that perfectly sums up the situation! Comparisons at that time were limited to known living animals, and naturalists had a hard time identifying fossils, often concluding them to be either relics of animals that had drowned in the biblical Flood or species that had not yet been identified. Was the "Maastricht unknown" a fish? A crocodile? A sperm whale? Hypotheses flourished, but no good match could be made. Naturalists were therefore stumped. (But let us not rush to cast aspersions on the scholars of the period, because the fields of geology and paleontology had not been established.)

Those jaws had been discovered two decades earlier, in quarries on the outskirts of Maastricht (fig. 2.83), during the excavation of freestone, a sedimentary rock from the Maastrichtian stage (ca. 70 Ma), which had been used since Roman times as a cut stone. This was how the quarry workers discovered some fossils, mostly of invertebrates (ammonites, bivalves, sea urchins, etc.), but occasionally of vertebrates (like the jaws we just mentioned). This was not the first time that the bones of this enigmatic animal had been found in these quarries. Large jaws comparable to those of the "unknown," on display today in the Teylers Museum in Haarlem (in the Netherlands), had been exhumed in 1770. It was, however, this unusual war trophy that made its way toward Paris in February 1795 and would enter the annals of history.

In the spring of 1795, the precious fossil became part of the collections of the Muséum national d'Histoire naturelle (MNHN), just opened in 1793 in Paris. And at exactly this time a young, ambitious, and brilliant naturalist joined the ranks of researchers at the MNHN. He was Georges Cuvier, 26 years old, originally from Montbéliard (which was then part of the German duchy of Wurttemberg) and just arrived from Normandy, where he had worked as a tutor for an aristocratic family during the Revolution (fig. 2.81). A happy circumstance therefore led to the meeting of what would become the most famous fossil and the person who would become the greatest naturalist of that age.

Already from his very beginnings at the MNHN, Cuvier launched into a gigantic and ambitious enterprise: to study the bones and the skeletons of all the vertebrates that crossed his path, whether they be living or fossil forms, and compare them to one another. He thus inaugurated a new method of study, which became known as comparative anatomy, an approach that would contribute to the

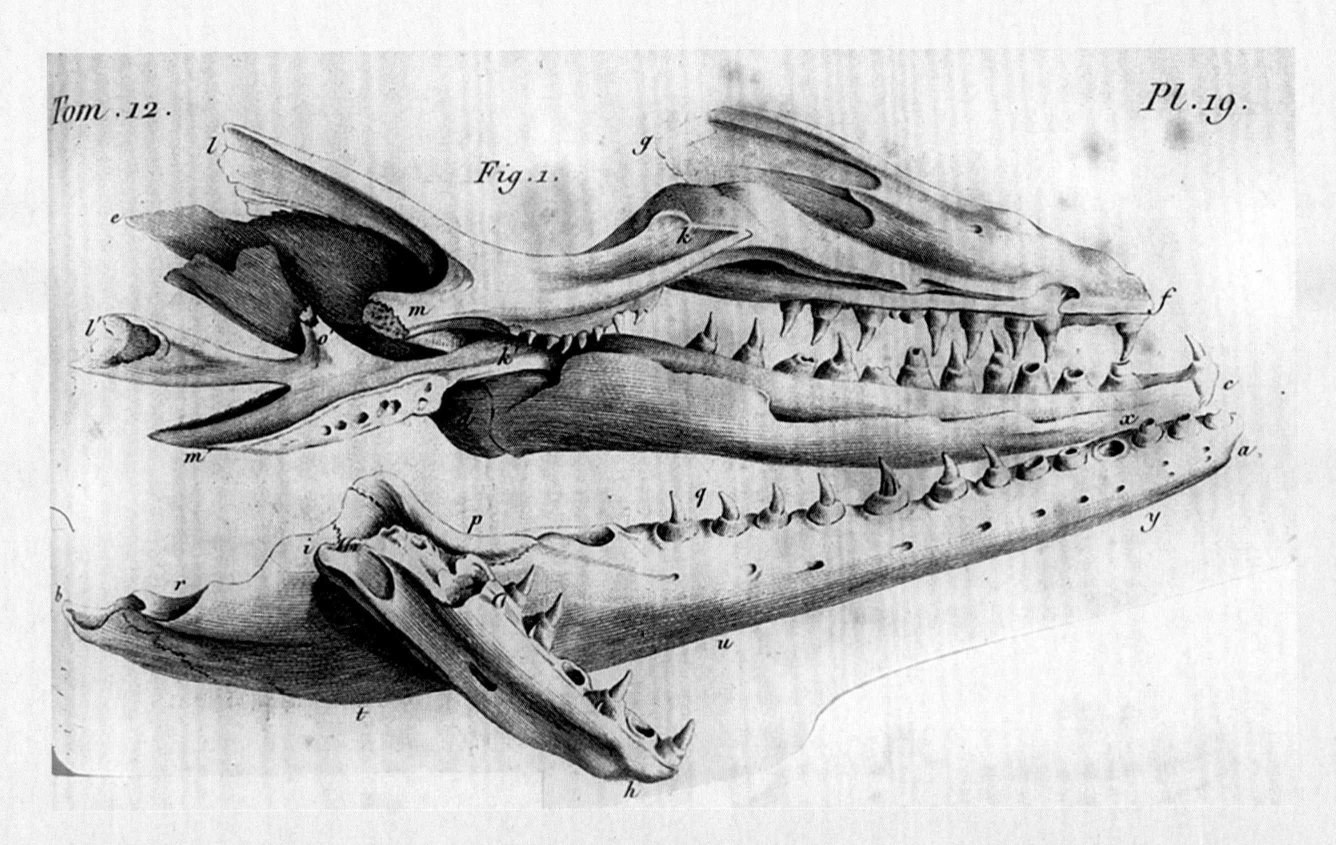

▲ Fig. 2.82. A plate from Georges Cuvier's 1808 article "Sur le Grand Animal Fossile des Carrières de Maastricht," in which he described the animal for the first time and determined its kinship relationships. This plate was the first rigorous scientific drawing of a mosasaur skull and reveals a set of characteristics typical of squamates.

foundations of paleontology itself. The principle of comparative anatomy is simple, as suggested by the name: its goal is to compare the structure and appearance (form) of the organs of different species, to distinguish **homologies** from **analogies**. Cuvier, moreover, had observed that within an animal, all the parts were connected. From the presence of one part, he could deduce the presence or absence of the others. This is a very practical way of working with incomplete fossils. Cuvier rightly boasted of being able to reconstitute an animal starting from a single bone.

After having studied several terrestrial mammals from the Quaternary period, such as the mammoth and the giant sloth from South America, then those from the Upper Eocene in the Montmartre gypsum quarries, including a famous opossum, Cuvier turned his attention to fossil marine reptiles. He studied the thalattosuchian crocodiles from the Jurassic in Normandy, but it was his masterful study of the "Maastricht unknown" that would shine the light of a new day on the history of a nascent paleontology and, more specifically, that of the large reptiles of the Mesozoic. This work of his constitutes the first precise anatomical description of a mosasaur, detailed and correctly illustrated (fig. 2.82), even though Cuvier, as would frequently be the case, gave it neither a scientific nor a species name, being satisfied with calling it the "Large fossil animal from the Maastricht quarries." This work also constitutes the first description of its nature—in other words, of a lizard closely related to today's monitor lizards, only giant, marine, and extinct! This interpretation was far more advanced than those previously mentioned (i.e., that the jaws were those of a fish, crocodile, or sperm whale). Cuvier had discussed it in some correspondence with a Dutch colleague, Adriaan Camper, who had reached the same conclusion as Cuvier. But it was Cuvier's description that became the go-to reference and is the one that is remembered.

First of all, Cuvier noticed that the jaws contained a number of bones. In that regard, they were comparable to those of reptiles and not to those of fish or mammals. Looking at reptiles, then, he observed that, their large size notwithstanding, the general shape of the jaws and the arrangement of the bones was closer to that of lizards, especially monitor lizards, than that of crocodiles. And finally, he deduced the animal's marine nature, partly by chance: the

▲ Fig. 2.83. The ancient quarries of Mount Saint Peter, on the outskirts of Maastricht (the Netherlands). It was in one of these underground tunnels, around 1774, that the first mosasaur to be described and named was discovered.

skull having been found in marine deposits, he concluded the animal must have lived in the sea. At the time, it was common practice to assume that an organism was necessarily fossilized in the environment it had lived in.

The Maastricht unknown was no longer unknown! Although it would not receive its scientific name of *Mosasaurus hoffmanni*—a reference both to the "lizard of the Mosa" (the Latin name of the river that runs through Maastricht—in Dutch, the Maas) and to the fossil's first owner, J. L. Hoffmann, in Maastricht—for another 20 years, it was the first mosasaur to be named, and it is therefore on it that the family name of Mosasauridae rests. It is also the largest mosasaur ever discovered and one of those from closest in time to the great Cretaceous/Paleogene extinction (see chapter 6, p. 179). More than two centuries after it was unearthed, this precious fossil still dominates the Paleontology Gallery of the MNHN, while a full-sized reconstruction is exhibited at the Natuurhistorisch Museum Maastricht, in the Netherlands (fig. 2.84).

After the study of the terrestrial mammals from the Quaternary system and Eocene series, Cuvier's study of the marine reptiles of the Mesozoic played an important role in his work, allowing him to continue to buttress his theory of catastrophism. Although mammoths lived in relatively recent times, the mammals from the Montmartre gypsum, the Maastricht mosasaur, and finally the thalattosuchians from Normandy had emerged from pasts that were increasingly remote. And they were all extinct! These fossils were therefore witnesses to worlds that had disappeared, inhabited by species greatly different from known living ones, and which had existed during different periods of Earth's history. Moreover, these species had been "lost"—they were therefore fixed and no longer evolved over the course of time—decimated by a series of universal catastrophes that Cuvier referred to as "revolutions of the globe." Although Cuvier might have been mistaken regarding the fixity of species, he had nevertheless arrived at an accurate perspective regarding the cyclical nature of the great extinctions that punctuate the history of life on Earth.

In conclusion, it is likely that Cuvier's canonical work on the Maastricht mosasaur influenced the image that the generations of paleontologists who followed had of the various groups of large Mesozoic reptiles—including and above all the dinosaurs—all portrayed in the form of "gigantic saurians," an undeniable reference to Cuvier's "lizard."

▼ Fig. 2.84. *Mosasaurus hoffmanni*, reconstructed skeleton on exhibit at the Natuurhistorisch Museum Maastricht.

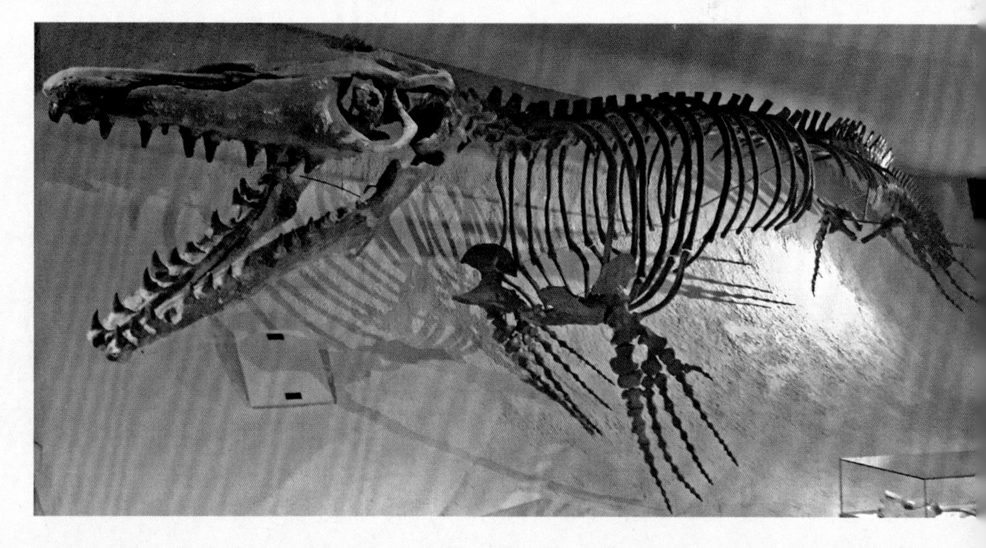

CHAPTER 3

THE PALEOZOIC ERA

First page of the score for "The Fossils" from the *Carnaval des Animaux* by Camille Saint-Saëns (1886).

The Late Paleozoic (300–252 Ma): "A Swimming Practice Center" for Reptiles

It Was Like an Apparition … and That Was It …

These two phrases—which provide the framework for the romantic adventures of the two heroes in Gustave Flaubert's *L'education sentimentale*—could be used to sum up the history of the conquest of the aquatic environment by reptiles at the dawn of the Mesozoic: a very promising "romance," but one that was ultimately never concluded. Three distinct and independent attempts had been made over the previous 50 million years, attempts that led nowhere, since these early adventurers have no close relations either with one another or with any of the groups of reptiles that would cast off their moorings several million years later, at the beginning of the Triassic.

▼ Fig. 3.1. *Spinoaequalis*, the first aquatic reptile, from the Upper Carboniferous in Kansas (United States).

Spinoaequalis, a Pioneer

This history therefore begins slowly but surely, more than 300 million years ago, during the Late Carboniferous. It was at that time, in what today is Kansas (but what then was the middle of Pangaea), that a small animal about 30 centimeters long, a true underdog, became the first aquatic reptile on Earth. *Spinoaequalis* (fig. 3.1) bears witness to the precocious diversification, both systematic and ecological, of one of the most flourishing groups, the diapsids, only a short time after the appearance of *Hylonomus* (the first reptile).

Several traits of *Spinoaequalis*'s skeleton, notably at the level of its tail, reveal a preference for water. Its caudal vertebrae were almost all devoid of transverse processes, while its neural and haemal spines were straight and very long. This allowed *Spinoaequalis* to both minimize its tail's

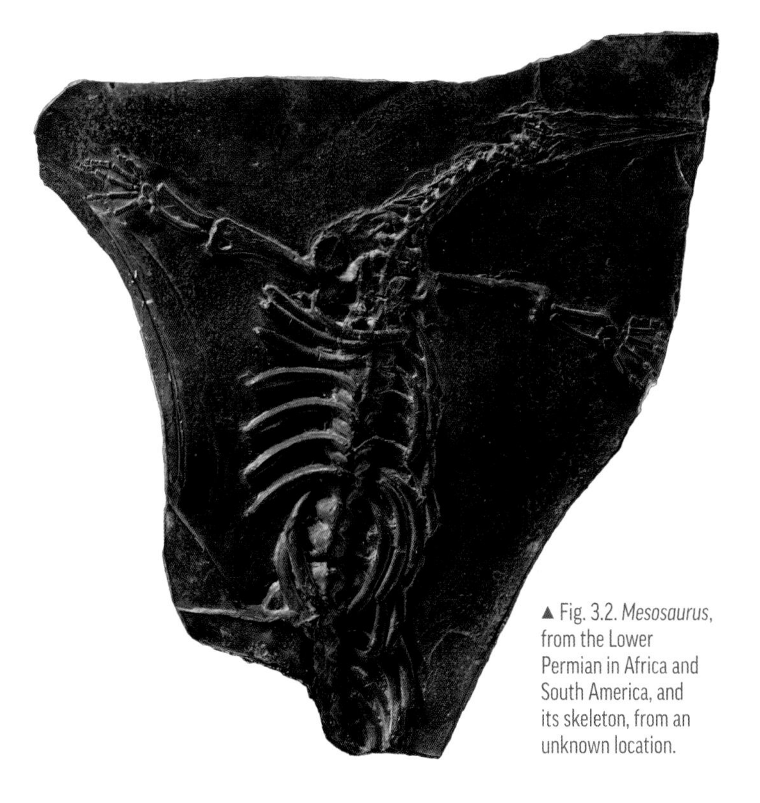

▲ Fig. 3.2. *Mesosaurus*, from the Lower Permian in Africa and South America, and its skeleton, from an unknown location.

▲► Fig. 3.3. *Claudiosaurus*, from the Permian in Madagascar, and its skeleton, from the Muséum national d'Histoire naturelle (Paris, France).

up-and-down movement and strengthen its side-to-side movement. *Spinoaequalis*'s tail therefore was high, flexible, and compressed laterally, making it suited to propulsion by means of lateral undulation, typical of an animal that has adapted to an aquatic way of life. Its limbs point to a terrestrial mode of life, and yet its state and location of preservation seems to point to adaptation to an aquatic environment. All this seems to suggest that *Spinoaequalis* was amphibious.

The Mesosaurs, Second Attempt

Several tens of millions of years later (290–270 Ma), an eternity for us but the blink of an eye on the geologic time scale, the mesosaurs—not to be confused with the mosasaurs!—first appeared and, in their turn, revealed some clear predispositions to an aquatic mode of life. Known only from the Lower Permian, these three genera, the best known of which is *Mesosaurus* (fig. 3.2), are related to neither *Spinoaequalis* nor the families of Mesozoic marine reptiles.

Mesosaurs have traditionally been considered anapsids, but the discovery of fossils exhibiting an inferior temporal fossa has revived the debate about their phylogenetic affinities. For certain experts, the presence of this inferior fossa points to mesosaurs being synapsids; for others, it could be a characteristic common to all amniotes! Whichever is the case, since 1999 the mesosaurs have been considered the sister-group to the other parareptiles ... and, several anatomical and phylogenetic revisions notwithstanding, this position seems stable.

Mesosaurs were of modest size, from several dozen centimeters to just under a meter in length. They had a hydrodynamic, torpedo-shaped body; a skull with a long, slender muzzle and numerous narrow, pointed teeth; webbed feet (as proven by fossil skin prints), pachyostotic ribs (thicker and denser than normal; see "The Secrets of Bone," p. 129); and a long, laterally compressed tail. All these attributes represent typical adaptations to an aquatic way of life. Mesosaurs' mode of reproduction remains uncertain, because, while possible embryos preserved in the abdomen of adult mesosaurs suggests they gave birth to live young (as aquatic reptiles frequently did), an isolated egg, which suggests **oviparity**, has been found too.

The small crustaceans preserved in the abdominal cavity of some specimens seem to indicate what mesosaurs ate. Although mesosaurs were long considered to have been tied to a shallow coastal environment, study of the paleoenvironment of Uruguayan deposits has yielded numerous mesosaur fossils and suggests that they frequented a hypersaline environment that was periodically isolated from the sea. Mesosaurs were just as certainly not cruisers of the open seas! Yet why, then, have they been found in regions as far apart as the western coast of Africa and the eastern coast of South America? We will return to the issue later. But let us first finish the last chapter of our Paleozoic history.

Claudiosaurus* and *Hovasaurus

Another several tens of millions of years later—still an eternity for us—in the Late Permian, about 255 million years ago, once again there arose some water-loving small reptiles. *Claudiosaurus* and *Hovasaurus* were discovered in Madagascar, which, at the time, was on the edge of Pangaea, attached to Africa. Both reptiles were about 50 to 60 centimeters long. Their long tail, flattened laterally, is yet another indication of aquatic tendencies. Despite its relatively long neck and a small head (fig. 3.3), *Claudiosaurus* was stockier than *Hovasaurus*, which somewhat resembled a lizard.

Just as in the previous cases, these two small aquatic reptiles have no kinship relation with the preceding aquatic reptiles or with those that later populated the seas. Let us open a small "aerial parenthesis" at this point: in these same Malagasy deposits are fossils of some small reptiles that attempted to overcome a very different challenge: gravity! Yes, reptiles were already beginning their conquest of the skies. During the Mesozoic, they would rule in all three environments—land, water, and air—which is why this era is known as "the age of reptiles." We should nevertheless note that, while Earth's oceans would be occupied by roughly a dozen orders of reptiles, only two reptilian orders, with rare exceptions, would share the vastness of the sky: the pterosaurs and the avian dinosaurs (the birds).

II
Mesosaurus: Alfred Wegener and Continental Drift

► Fig. 3.4. Alfred Wegener, father of the theory of continental drift.

Although the idea was "in the air" at the time, it was German meteorologist Alfred Wegener (1885–1930) (fig. 3.4) who, based on a solid multidisciplinary scientific argument, first elaborated the theory of "continental drift" (Kontinentalverschiebungen), in a scientific article in 1912 and then an entire work on the subject in 1915.

At the time, the commonly held belief was that the earth's configuration had always been the same, continents and oceans having always occupied the locations we know them to be in today. How, then, to explain the presence of identical terrestrial faunas and floras (modern as well as ancient) in different parts of the globe? For example, how is it that the common garden snail (fig. 3.5) is known in both Europe and North America, continents separated by the Atlantic Ocean, an impassable barrier for this little gastropod? The explanation offered by many, including renowned scientists such as Charles Darwin and Alfred R. Wallace, was migration enabled by flooding or by "continental bridges."

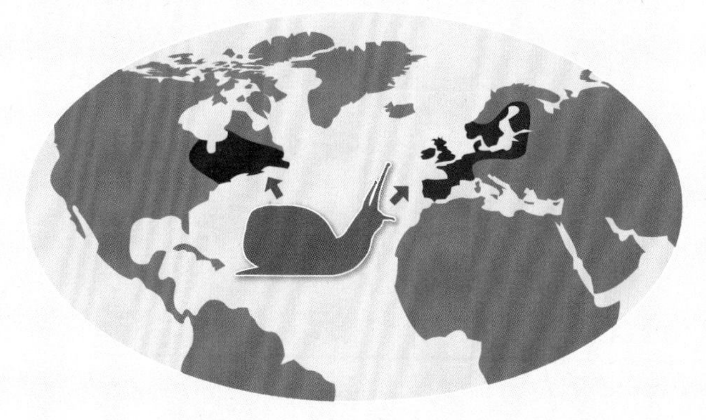

▲ 3.5. The current distribution of the garden snail.

Once upon a Time, There Was a German Meteorologist ...

Wegener proposed a particularly innovative explanation for these strange distributions. While looking at a world map in 1910, he saw what many before him had seen: the coastlines of South America and Africa fit together almost perfectly. The following question came to him: had these two continents at one point been joined? After gathering an immense bundle of scientific evidence (geodesic, geophysical, geological, climatological, biological, and paleontological), he imagined an opera in three acts (fig. 3.6):

1. At the end of the Paleozoic, about 260 million years ago, all of today's continents were joined in a single continental mass, called Pangaea, set in an immense ocean, Panthalassa; an outgrowth of this ocean, the Tethys, cut into Pangaea in the east, like a wedge.
2. At the beginning of the Triassic, about 250 million years ago, Pangaea began to split apart, giving rise to Laurasia (Asia, Europe, and North America) in the north and to Gondwana (Africa, Oceania, Antarctica, South America, India, and the Arabian Peninsula) in the south; this is also how some ocean inlets—embryos of today's oceans, specifically the Atlantic Ocean—began to open.
3. From this point on, fragmenting continued, following perpetual movement, and continents continued to drift apart, changing the face of the earth in each age. The continents and the oceans we are familiar with are therefore nothing more than a stage in this process, and millions of years from now they will occupy very different locations.

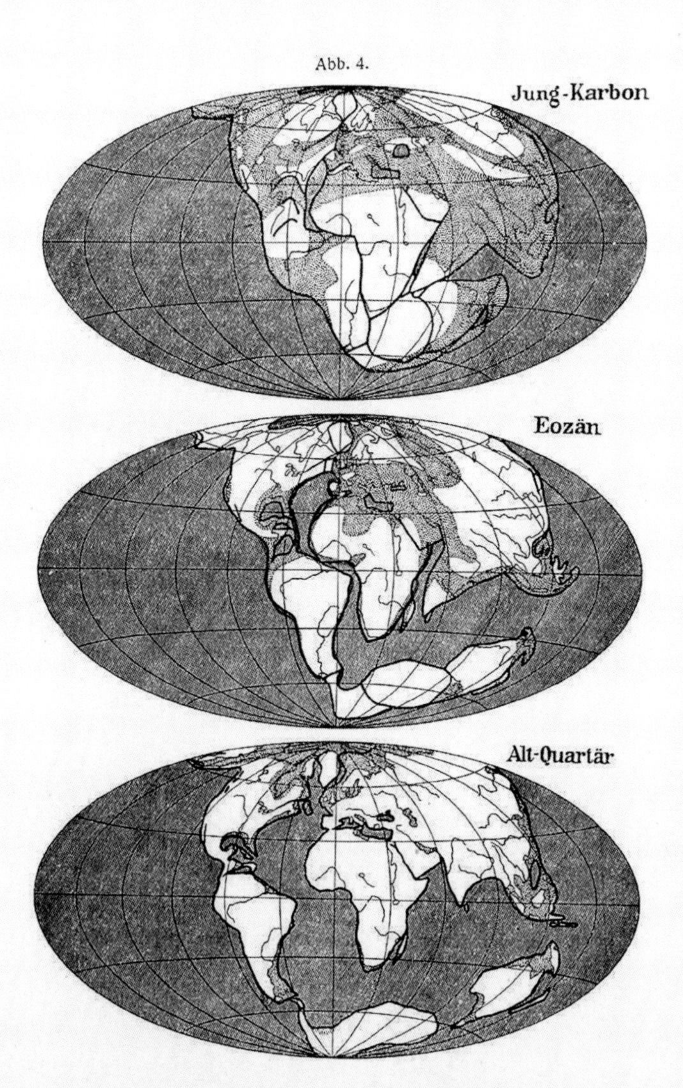

▲ 6. Fig. 3.6. Famous maps published by Wegener in his 1915 work. They show the breaking apart of Pangaea and the drift of continents that resulted. There is a time lag regarding the ages as compared to modern maps: the map of Earth during the Eocene (middle) would, to today's geologists, correspond to a map of Earth during the Early Cretaceous.

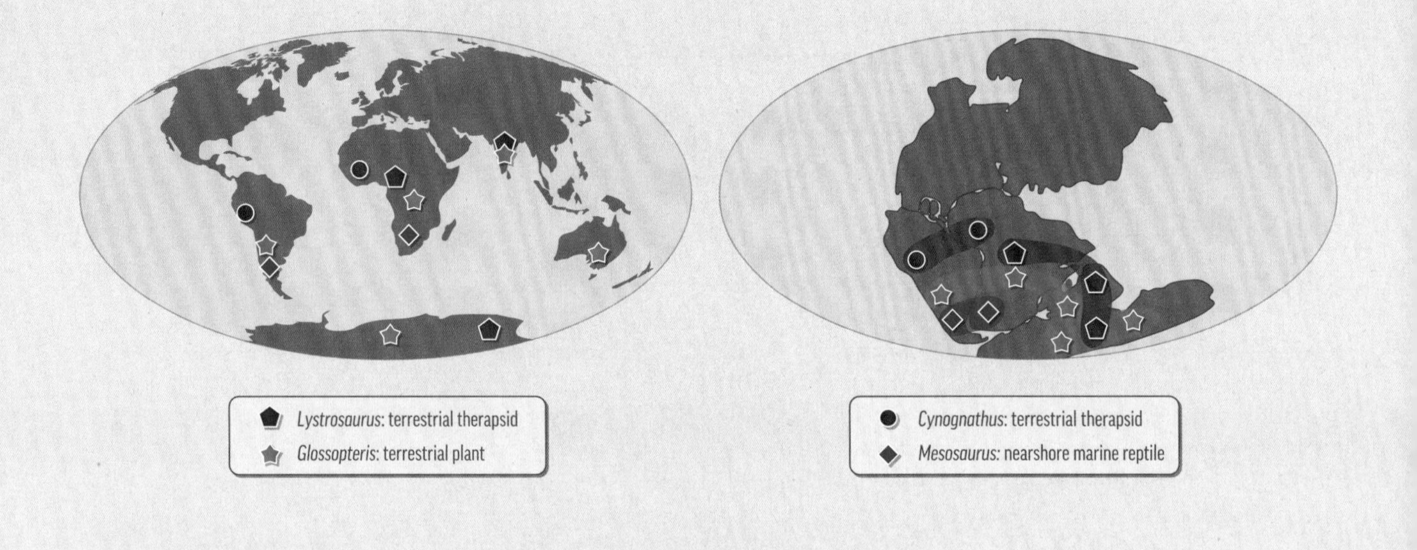

▲ Fig. 3.7. The distribution of some continental (or nearshore marine) fossils of plants and animals in the Permian period, represented on a current map and on a map of Pangaea as imagined by Wegener: their areas of distribution seem to have been contiguous.

▼ Fig. 3.8. Areas where *Mesosaurus* fossils were found.

Mesosaurus, One of the Heroes of the Wegenerian "Trilogy"

To support his theory, Wegener used several fossil taxa dating from 260 to 240 million years ago: the *Glossopteris* plant, the therapsids *Cynognathus* and *Lystrosaurus*, and the reptile *Mesosaurus*, all found on both sides of the Atlantic, in South America and in South Africa (fig. 3.7). How could continental (or coastal marine) organisms have colonized geographic areas so distant from one another? The existence of a mega-continent during the Permian allowed Wegener to propose a coherent distribution for all these taxa and, thereby, a much simpler hypothesis than "continental bridges."

The small aquatic reptile *Mesosaurus*, for instance, would have been incapable of crossing the Atlantic if the ocean had been as large during the Permian as it is today (fig. 3.8). If we mark the locations where fossils of these creatures have been found on a map of Pangaea, we see that they are concentrated in one area, called the "sea of the mesosaurs," a shallow marine expanse south of Gondwana, between the tip of Africa and the tip of South America. The oceans that emerged from the fracturing of Pangaea and were serious barriers for continental animals were handy avenues for the dispersal of marine ones. It has been shown that during the Mesozoic, on the heels of Pangaea's fragmenting, the different groups of marine reptiles spread out into the oceans and the inlets that the breakup created. What separated the continental animals brought the marine ones together!

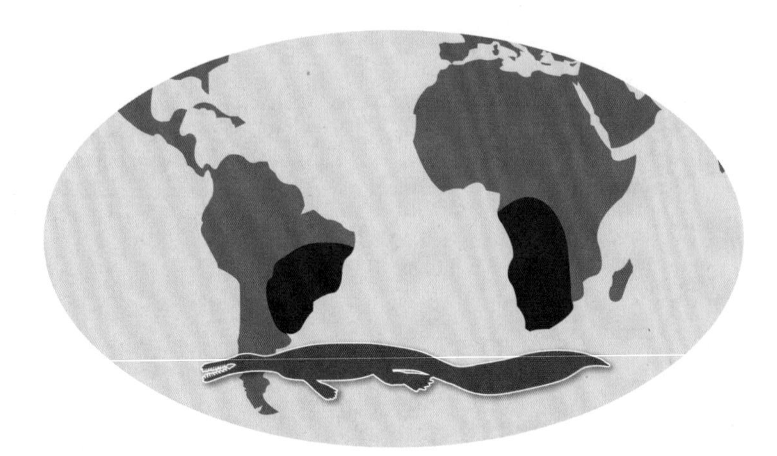

"We Are Missing a Newton to Explain the Enigma of the Mechanism of Continental Drift"

This is how Alfred Wegener expresses himself, with great modesty, in his 1915 work.

And rightly so, since his theory did not propose any mechanism to explain continental drift, and therefore for a long time it was not given the recognition it deserved. But it was certainly not ignored ... in fact, it is a veritable paradigm within the earth sciences: after Darwin's theory of evolution, Wegener's theory of continental drift is the theory that has mobilized scientists in different disciplines the most. In the 1950s and 1960s, it was the study of paleomagnetism, as well as the exploration of the ocean floors, that finally contributed some elements of the answer to what drives the movement of the continents. So, in 1968, the theory of tectonic plates was born, which replaced—by integrating it—the theory of continental drift. Finally, after 50 years, Wegener's hypothesis was recognized and accepted by the scientific community.

In its turn, however, the theory of tectonic plates does not resolve all questions. The latest discoveries in the field have shaken up even well-established concepts, like that of oceanic expansion starting from ocean ridges and the "conveyor belt" model proposed by American geologist Harry Hess. According to this model, which became the classical model starting in the early 1960s, solid tectonic plates move about on top of a mantle of viscous rock, as a result of convective movement within the mantle itself. Since Earth's volume remains identical, the expansion of the ocean floors at the level of ocean ridges is compensated for by subductions (the plunging of a dense oceanic plate beneath a lighter continental plate) or by collisions between continents (between continental plates of the same density; for example, the collision that created the Himalayas). But after about 20 years, a second model emerged, in which subductions are the principal motor for the shifting of tectonic plates. To simplify the model somewhat, the cooling of the oceanic plates increases their density, and once they move closer to the lighter continental plates, they plunge beneath them, owing to subduction. But in this model, it is only their weight, which pulls oceanic plates downward, that is the cause of the displacements on the whole; the plates are not influenced by the convective movements of the mantle, which are too chaotic and too slow to lead to these kinds of displacements. Experts could not reach agreement on which model is more correct until a detailed simulation was run, using supercomputers continuously for several months. The result, surprisingly, was that both models are right ... pretty much depending on the period and the location. It seems that subductions are currently responsible for about two-thirds of tectonic movements, while currents in the mantle may have been the determining factor in the case of collisions or of the fragmenting of continents. Wegener can rest assured: we are perhaps not that far from having found our Newton ... a Newton supercomputer?

CHAPTER 4

THE TRIASSIC PERIOD

A Hungarian postage stamp depicting the placodont *Placochelys placodonta.*

I
A Period of Great Changes

The Triassic (252–201 Ma), the first period of the Mesozoic era, was defined in 1834 by German geologist Friedrich August von Alberti. Its name (from the Latin *trias*, "triad") refers to the three geologic layers, or lithostratigraphic units, that characterize this time period in central Europe: the Bundsandstein is composed essentially of variegated sandstone, the Muschelkalk is a shelly limestone, and the Keuper contains salt-bearing rocks.

At the very beginning of the Triassic, the marine world, probably even more than the terrestrial environment, had been devastated by the Permian/Triassic crisis (see chapter 1, p. 16). Only a small percentage of the species on Earth in the Permian had survived. The Triassic is therefore a particularly captivating period for paleontologists interested in the various stages of the recovery and diversification of life that occurs after such profound environmental changes.

The ecosystems at the very beginning of the Triassic were composed mainly of generalist species, capable of resisting significant environmental stress, species we might call "disaster-proof." Such species "profit," in some manner, by moving into the

▼ Fig. 4.1. A Middle Triassic ocean scene, Switzerland. On the right are two marine reptiles, *Nothosaurus* (above) and *Placodus* (below). Two hybodont sharks patrol close to the bottom (on the left), beneath a shoal of *Gyrolepis*.

Top Predator: *Nothosaurus*

Name: "fake lizard"
Classification: Sauropterygia, Eosauropterygia, Nothosauroidea, Nothosauridae
Lived: Middle Triassic to Late Triassic (Anisian–Carnian, 240–225 Ma)
Known range: Europe, North Africa, China
Overall length: up to 4 meters
Diet: fish

The genus *Nothosaurus* includes around ten species of very variable size. *N. juvenilis*, the smallest, is known from a single skull about 13 centimeters long. The skull of the largest species, *N. giganteus*, was five times as long, in an animal that could reach 4 meters in length.

Nothosaurs share a common ancestor with plesiosaurs and belong to the group Eosauropterygia. While reptiles in both groups exhibit adaptations characteristic of aquatic life, the proportions of nothosaurs' forelimbs and hind limbs were still relatively close to those of terrestrial animals, with the lower bones of both the front limbs (radius/ulna) and the hind limbs (tibia/fibula) all being long. Yet these animals must still have spent the better parts of their lives in water.

ecological niches of those that have perished and then proliferating until competition starts to increase. The diversification of the disaster-proof species, which happens once the postcrisis environmental conditions become less stressful, is nevertheless not a sign that the ecosystem has recovered. Recovery is usually considered to have begun once the disaster-proof species have been replaced by more specialized varieties that will support the ecosystem over the long term and will fill the empty niches in many living environments. At the beginning of the Triassic, creatures generally diversified starting from communities in areas that had been relatively protected during the crisis, areas that therefore functioned as nurseries. This recovery, up to the development of new food chains, took 5 to 10 million years. For some students of the history of life on Earth, the Triassic is the time that saw the emergence of modern faunas, and, in effect, this is indeed when numerous predators (marine reptiles but also invertebrate predators) exhibiting novel morphologies and adaptations, as well as previously unknown ways of life, first appeared.

As far as the marine reptiles are concerned, a veritable horde of predators (ichthyosaurs, sauropterygians, thalattosaurs, and archosauromorphs), with no kinship relations to the several pioneers that took to the waters during the Permian (see chapter 3, p. 101), entered the scene at the beginning of the Triassic. For some of these groups, the adventure would last no longer than the Triassic itself (which did span 50 million years!), but for others, such as the sauropterygians and the ichthyosaurs, it would extend for much longer. Contrary to previous belief, these lineages of marine reptiles were neither exclusively small nor entirely restricted to the coastal areas of Pangaea. New research has demonstrated that (a) only a couple of million years after the Permian/Triassic crisis, numerous entirely new food chains already existed, with reptiles/fish replacing amphibians/fish as the dominant group at the top, and (b) as early as the beginning of the Middle Triassic, marine reptiles were already great voyagers, since their fossilized remains have been found almost all around the globe in deposits from that time.

The ichthyosaurs, which left a very rich fossil record since the beginning of the Triassic, are without a doubt the group that teaches us in the most significant ways about these phenomena. For instance, while *Chaohusaurus* didn't exceed 70 centimeters in length, and *Grippia* did not exceed a meter, *Utatsusaurus* was already flirting with 3 meters in length. *Thalattoarchon* appeared only 4 million years after the Permian/Triassic crisis but already reached almost 9 meters in length; it was the first marine mega-predator of the Mesozoic. Fossils have been found of shastasaurids, a family that includes the largest ichthyosaurs of all, dating from around this same time. All these Lower Triassic taxa were widely distributed in northern Pangaea and have been found from British Columbia to Thailand, by way of Norway and China. Ichthyosaurs' range even grew at the beginning of the Middle Triassic, when they invaded the western coastline of Pangaea and the entire Tethys. By that time, some ichthyosaurs already had begun to exhibit the typical body type that would ensure their success for the next 140 million years, and they were already giving birth to live young! (see chapter 5, p. 143).

In conclusion, the ocean plunge taken by reptiles starting at the beginning of the Triassic represents one of the major adaptive radiations in the history of life on Earth. It happened much more rapidly than very recently believed: only several million years after being devastated by the greatest extinction of all time, the marine environment was once again teeming with life, this time including numerous and extremely varied groups of ocean-crossing reptiles. A stunning evolutionary success!

II Monte San Giorgio

Geographic and Stratigraphic Context

In the Southern Alps, on the border between the canton of Ticino in Switzerland and the Lombardy region of Italy, near Lake Lugano, is a remarkable mountain called Monte San Giorgio. Monte San Giorgio's sedimentary rocks, laid down over a period of more than 100 million years, were thrust up to their current elevation more than 1,000 meters above sea level during the formation of these mountains, in the Paleogene period. The sedimentary deposits, angled by their uplifting, form a continuous geologic record (a succession) stretching from the Permian to the Jurassic, about 600 meters thick (fig. 4.2). In 2003 Monte San Giorgio was classified as a UNESCO World Heritage Site because of the hundreds of important fossils that have been extracted from its remarkable deposits.

In the middle layers of this succession (fig. 4.3), both stages of the Middle Triassic, the Anisian and the Ladinian, are represented. During the Anisian age (247–242 Ma), the area was part of a river delta, and the corresponding geologic sequence, the Bellano formation, has not revealed any fossils of marine reptiles. But a rise in sea level at the very end of the Anisian led, during the Ladinian (242–237 Ma), to the establishment of a shallow sea, and the area developed into a lagoon 30–100 meters deep. At the time, the region was situated at about 18 degrees north latitude and was probably influenced by a subtropical monsoon climate.

The underlying Besano formation (or Grenzbitumenzone) consists of alternating layers of black schist and dolomite. It is covered by the deposits from the Ladinian, which are composed of dolomite overlaid by limestone interspersed with four fossil-rich layers: three layers of black schist—called, respectively, Cava Inferiore, Cava Superiore, and Cassina—and a layer of clayey limestone that forms the Kalkschieferzone.

A Little History

The Besano formation contains the oldest fossil-rich sediments in this region, and it is the one that houses the most spectacular fossils. This sedimentary formation has been studied and exploited since the first half of the nineteenth century.

The first scientific finds date from 1847 and were mostly of fossil fish. Then, during the second half of the nineteenth century and the

◄ Fig. 4.2. A Middle Triassic deposit from Monte San Giorgio, which straddles the border between Switzerland and Italy.

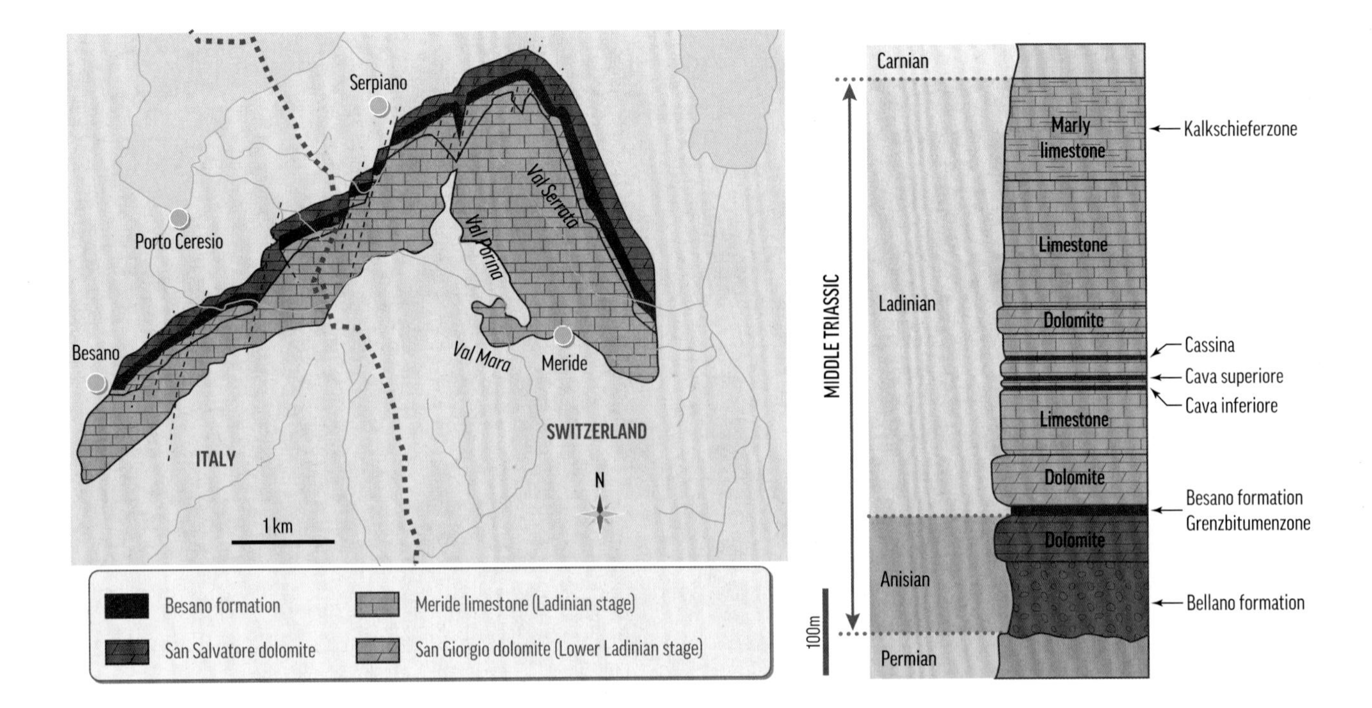

▲ Fig. 4.3. On the left: The geographic extent of the fossil-rich formations of the Monte San Giorgio deposit. On the right: the stratigraphic column of the Triassic series, showing the main fossil-rich formations.

first half of the twentieth century, these layers of bituminous rock (containing hydrocarbons) were exploited industrially for the production of ichthammol-based pharmaceuticals used to treat rheumatism and other conditions. The more recent fossil-rich layers (Cava Inferiore, Cava Superiore, Cassina, and Kalkschieferzone) have provided less spectacular specimens of fish, small reptiles, and embryos of reptiles, in addition to some insects.

The first fossils of marine reptiles at Monte San Giorgio were discovered in 1863 in the Besano formation, on the Italian side of the mountain. *Tanystropheus* and *Mixosaurus* were described in 1886. Unfortunately, in 1943, some of these excavated fossils, at the natural history museum in nearby Milan, were destroyed during the Allied bombardment. However, these incalculable losses were partially compensated for by the marine reptiles unearthed on the Swiss side starting in 1924. Among them were some specimens of *Cyamodus* (a placodont), *Tanystropheus* (fig. 4.4), and dozens of specimens of the ichthyosaur *Mixosaurus*. But this is not the whole story. In the Middle Triassic, since the lagoon was not far from dry land, sometimes the remains of terrestrial animals—such as the almost 3-meter-long carnivorous archosaur *Ticinosuchus*—ended up in these waters and sank to the bottom, where they fossilized together with the remains of marine animals. Add to this bounty the fossil insects and fossil vegetation unearthed here, and Monte San Giorgio provides us with a rare glimpse of continental biodiversity on the shores of an inland Triassic sea.

Thanks, Bacteria!

The black schists of the Middle Triassic are fine and laminated (thinly layered) sediments, which indicates that the setting they were deposited in was a calm one, with a complete absence of **bioturbation**. The thin layers may be the result of bacterial mats that developed on the ocean floors, with most of the organic matter contained in these schists having belonged to these bacteria.

The vertebrate fossils in these schists contain articulated (i.e., connected at the joints, as in life) and mostly complete skeletons, indicating that the remains of most of these fish and marine reptiles descended to the seabed relatively quickly, before scavengers could dismember them. Decomposition must have started only once the remains were on the bottom but must have been stopped rapidly by the bacterial mats.

Several articulated specimens of marine reptiles at Monte San Giorgio have been found preserved in an unusual fashion: belly-side-up. The most plausible explanation for this is that, after death, the accumulation of gases from putrefaction in the abdominal cavity created buoyancy, lifting the body to the surface, where it floated on its back. Once the body sank to the bottom again, it maintained this position. This phenomenon is not unique to Monte San Giorgio, but it is not easily observable in deposits where the preservation is not as fine or where animals were often partially dismembered before they could be covered by sediment and begin to fossilize.

Several articulated specimens of marine reptiles have, in an unusual fashion, been found preserved belly-side-up.

Noteworthy Biodiversity

In this deposit, the fossils are very diverse, numerous, and often complete and articulated. Nevertheless, the identification of certain groups, such as the sharks, depends mainly on the study of the remains of teeth and spine, because these animals' cartilaginous skeleton decomposes rapidly and thus fossilizes only very rarely. And yet several complete specimens of sharks have been found. At least five species of sharks, as well as a great number of fossil fish belonging to different groups of **sarcopterygians** and actinopterygians (e.g., *Saurichthys*, *Peltopleurus*, and *Archaeosemionotus*), have been identified. As far as marine reptiles are concerned, fossils belonging to four large groups have been identified: ichthyosaurs, sauropterygians, thalattosaurs, and some enigmatic archosauromorphs.

Ichthyosaurs

Monte San Giorgio has yielded five genera of ichthyosaurs: *Mixosaurus*, *Besanosaurus*, *Cymbospondylus* (fig. 4.4), *Wimanius*, and *Mikadocephalus*. The most common remains are those of *Mixosaurus*, about 1.5 meters long. *Mixosaurus* has also been found in the United States (Nevada, Canada, and Alaska), in China (Yunnan and Guizhou), in Indonesia, and in Svalbard, which leads us to believe it was spread around the globe. Its morphology—between that of more primitive forms such as *Cymbospondylus* (6–10 meters long), which had a long and tapered body, and more evolved forms, which resembled sharks and dolphins—earned it its name of "mixed lizard." *Mixosaurus*, in effect, does not exhibit pronounced development of the dorsal and caudal fins. Moreover, although it exhibits its moderate hyperphalangy, its limbs had no more than five fingers (i.e., no hyperdactyly). Its bones do not display the increase in mass (see "The Secrets of Bone," p. 129) sometimes characteristic of coastal animals that swim relatively slowly and rarely venture

▼ Fig. 4.4. (Following pages) The Middle Triassic, Monte San Giorgio, on the border between Switzerland and Italy. **Above**, a beach scene. Two archosauromorphs: a *Tanystropheus*, with its immeasurably long neck (in the middle), and a *Macrocnemus* under the leaves (on the right). **Below**, an underwater scene: in the middle, two ichthyosaurs, the imposing *Cymbospondylus* chasing the small *Mixosaurus*. *Cyamodus*, an armored placodont, crosses paths with a thalattosaur, *Askeptosaurus* (in the background on the left). Two actinopterygians, *Saurichthys*, and, close to the rocks, two sauropterygians, *Neusticosaurus* (above and to the right), complete the scene.

Adaptations to the Aquatic Environment

Marine reptiles of the Mesozoic were descended from animals that lived on dry land. The terrestrial and aquatic environments are milieus with very different constraints: weight is an important element on dry land, much less so in the water, but the density and the resistance of the liquid element have a significant impact on movement. The reconquest of the aquatic environment, particularly the marine one, thus relied on the natural selection of numerous morphological transformations (a more hydrodynamic body, transformation of limbs into swimming paddles, etc.) as well as physiological ones (modifications of the sensory organs, a change in mode of reproduction, adaptations to withstand salinity and diving, etc.). These adaptations vary in importance depending on whether we are discussing coastal animals (which were still capable of moving about on dry land) or animals that were completely independent of the terrestrial environment. The marine reptiles of the Mesozoic therefore exhibit different degrees of adaptation to aquatic life.

The types of anatomical characteristics we see among tetrapods secondarily adapted to an aquatic mode of life were imposed by the functional constraints on effective swimming and predation, leading to modifications that provided better propulsion, maintenance of bodily position, and drag reduction.

Form of Swimming

Most of today's aquatic vertebrates move by contracting the muscles attached to their spine, first on one side and then on the other (axial locomotion), but different groups use different portions of their body to produce the propulsive force. These differences can be said to have created four forms of swimming, characterized by both different body types and different ways of moving (fig. 2.5):

1. Anguilliform swimming involves undulating the entire body, in the manner of a freshwater or a moray eel. Although this mode of propulsion is not very efficient because a lot of energy is wasted laterally, it allows for great maneuverability. Sea snakes swim in this way.
2. Subcarangiform swimming corresponds to an axial subundulation, where only the rear half of the body moves, while the front remains rigid, as seen in pike, crocodiles, and marine iguanas.
3. Carangiform swimming (axial suboscillation) involves only the rear third of the body. Animals that use this form of swimming today include blue sharks, sardines, and manatees.
4. Thunniform swimming corresponds to an axial oscillation, in which only the caudal (tail) fin moves. Think of tuna, swordfish, or white sharks.

Most tetrapods that have adapted to shallow aquatic environments exhibit an elongated body, a long tail that is compressed laterally, and limbs that, while perhaps smaller, have not been modified into swimming paddles. They are anguilliform swimmers. Since an elongated and flexible body allows for large, wide undulation and rapid acceleration but increases drag, animals that rely on anguilliform or subcarangiform swimming are mostly poor swimmers that do best to ambush their prey.

At the opposite end of the spectrum, thunniform swimmers are the fastest. Usually pursuit predators, they often exhibit a torpedo-shaped body and a symmetrical, crescent-shaped caudal fin, with a narrow point of attachment.

▼ Fig. 4.5. From left to right, the four swimming types and frames of movement associated with axial locomotion: anguilliform (in which three-quarters of the body changes shape and assists in propulsion), subcarangiform (in which the rear half of the body is used), carangiform (in which only the rear third of the body is put to work), and thunniform (in which only the caudal fin oscillates).

These characteristics reduce friction and increase stability at high speeds.

Those vertebrates that do not use axial locomotion—such as clownfish, seahorses, sea turtles, penguins, and sea lions—use paraxial locomotion instead. They use all other even (limbs) or uneven (dorsal fin) appendages, except for the tail.

Buoyancy Control

In water it is necessary to oppose buoyancy in order to dive and then remain in the deep. In animals that have returned to an aquatic environment, this problem is significantly increased by the presence of lungs. Oxygen reserves are even more important if the animal is a good free diver (this is the case especially for shallower depths; deeper divers store oxygen in their muscles), because the greater these reserves are, the longer the animal will be able to stay underwater. Tetrapods, consequently, have the tendency to naturally rise toward the surface, and the effort they must exert to maintain depth is therefore very significant.

Creatures that are active swimmers can accomplish this goal by means of their swimming style, just as we do when we move from the surface toward the bottom of a swimming pool. However, for the less active swimmers, such effort wastes too much energy. One solution for these latter creatures, then, is to increase their weight, in order to reduce their buoyancy and therefore enable themselves to descend more easily and to remain at depth. Numerous kinds of aquatic tetrapods today evolved structures, such as osteoderms in crocodiles and shells in turtles, that increase their body mass. In addition, some creatures resort to swallowing stones for ballast. These options notwithstanding, the most common strategy for countering buoyancy is to increase bone mass. This can be achieved by osteosclerosis (increased bone density), via pachyostosis (the addition of peripheral bone deposits), or even by a combination of the two (pachyosteosclerosis) (see "The Secrets of Bone," p. 129).

Attitude Control

Because its lungs (which are light and full of air) are toward the front of its body, a tetrapod's natural center of gravity is a considerable distance toward its rear. In the water, this has the unintended effect of making it hard to maintain a horizontal body position (fig. 4.6). The body is therefore generally on an incline, with the head higher. An increase in the mass of the anterior part of the body can shift the center of gravity toward the front, helping the animal stay level. An increase in bone mass at the front half of the body (generally in the area around the lungs) is the kind of specialization most frequently developed for this purpose. However, certain animals for which a permanent increase in bone mass would create a disadvantage use other means of controlling their bodily attitude.

An increase in anatomical density would obviously not work well in meeting the functional needs of hunters that rely on speed. A particularly weighty skeleton would reduce their maneuverability and hamper the swimming and acceleration capabilities on which they depend. They therefore control their buoyancy and their pitch angle in a dynamic fashion, by means of their active swimming style. It is thus possible to determine, based on their anatomy and the internal structure of their skeleton, which form of swimming (anguilliform, subcarangiform, carangiform, or thunniform) different Mesozoic marine reptiles used.

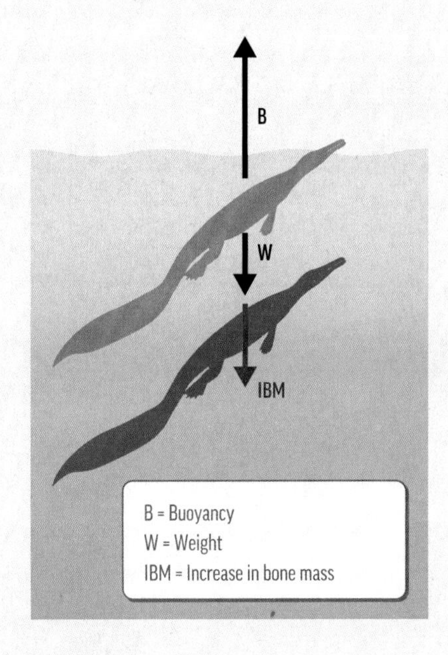

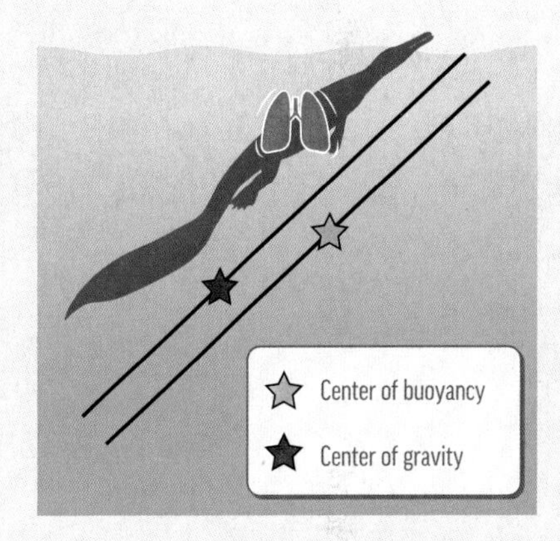

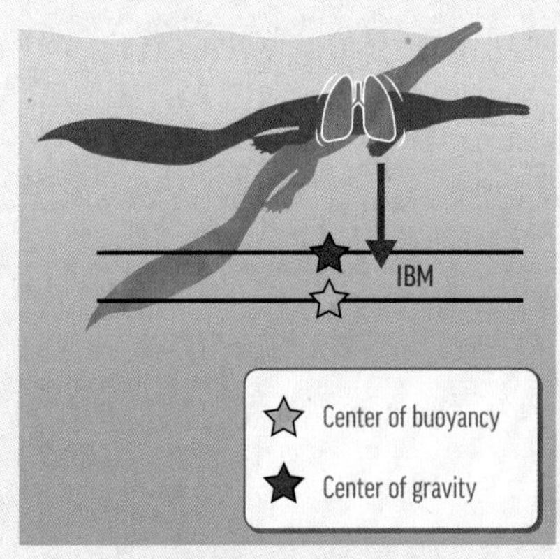

▲ Fig. 4.6. Control of body position (pitch angle).

into the open ocean. Monte San Giorgio has yielded *Mixosaurus* specimens pertaining to different stages of **ontogenetic** development, ranging from juveniles roughly half a meter long to adults measuring 1.5 to 2 meters. *Mixosaurus* was equipped with massive teeth in the rear part of its jaw, for crushing prey, as well as sharp conical teeth in the front.

In addition to *Mixosaurus*, other small ichthyosaurs, such as *Wimanius*, were evolving. *Wimanius*'s skull was about 25 centimeters long. *Mikadocephalus* was slightly more imposing, and its skull was twice as long as *Wimanius*'s. The genus received its name from the famous game of mikado (pick-up sticks) because the remains of the skull discovered at Monte San Giorgio were preserved, dispersed, and superposed on one another in a fashion similar to the sticks in the game.

Reigning over this ancient body of water, however, were some significantly larger ichthyosaurs: *Besanosaurus* and *Cymbospondylus*. *Besanosaurus*, discovered in 1993 on the Italian side of the mountain, was 5 or 6 meters from its snout to the tip of its lengthy tail, which contained close to 140 vertebrae. It lacked a dorsal fin, and it had a long, narrow muzzle, with very small teeth. *Cymbospondylus* (fig. 4.4) was characterized by an anguilliform body 6–10 meters long, devoid of both dorsal fin and bilobed tail. The fourth-largest known ichthyosaur, it probably was piscivorous. Following its description in 1868, for a long time it was included in the family Shastasauridae, but it is now considered to have been a much more primitive kind of ichthyosaur. Other places *Cymbospondylus* has been found are the United States (Nevada and California) and Germany.

About 5 centimeters long, the nearly complete skeleton of a *Neusticosaurus* embryo (a pachypleurosaur) is one of the smallest vertebrate fossils ever.

A Sauropterygian Paradise

Several hundred fossils of sauropterygians have been found in the Monte San Giorgio deposits. They belong to the pachypleurosaur *Neusticosaurus* (fig. 4.4), the nothosaurs *Lariosaurus* and *Ceresiosaurus*, and the placodonts *Cyamodus* (fig. 4.4) and *Paraplacodus*. The size variations among sauropterygians were very considerable. *Ceresiosaurus* could attain a length of 3 meters and essentially held the role of mega-predator in this body of water, while *Neusticosaurus* did not exceed 30 centimeters in length. *Neusticosaurus* fossils are very abundant, and many hundreds of specimens are cataloged in the Zurich Paleontological Institute. Among them is the fully articulated, nearly complete skeleton of an embryo. About 5 centimeters in length, it is one of the smallest vertebrate fossils ever! The tiny specimen was found rolled up, with its head near its tail, a posture not encountered among other specimens from the same deposit but one that is indeed quite typical for the embryos of vertebrates. *Lariosaurus* embryos were also discovered, and more specifically three of them were found agglutinated (clumped together). Neither the *Neusticosaurus* nor the *Lariosaurus* embryos were found in the abdominal cavity of an adult, yet researchers could discern no trace of a shell surrounding them. Paleontologists therefore wonder about these creatures' mode of reproduction: viviparous or oviparous?

The Enigmatic Thalattosaurs

Three genera of thalattosaurs have been found at Monte San Giorgio: *Clarazia*, about

1 meter long, equipped with teeth good for crushing, which it probably used to pulverize the shells of invertebrates; *Hescheleria*, with an anatomy and lifestyle that are poorly known; and *Askeptosaurus* (fig. 4.4), a piscivore that could attain a maximum length of 3 meters. *Askeptosaurus*'s skeleton retains certain characteristics of terrestrial animals—for instance, the carpal (wrist) and tarsal (ankle) bones are well ossified, indicating it probably undertook forays onto land. *Askeptosaurus*'s long neck may have helped it capture fast-moving fish.

Two Strange Archosauromorphs

The most enigmatic marine reptile discovered in these deposits is, without a doubt, *Tanystropheus*. Its name is derived from the Greek and signifies "long vertebrae." This genus of prolacertiform (a group of archosauromorphs) is known from only two or three species that have been discovered in different parts of the globe: Europe, China, and Israel. About 5 meters long, *Tanystropheus* had a barrel-shaped body, a small head, and a neck generally about twice the length of its trunk but sometimes as long as the entire rest of its body (fig. 4.4). That neck contained only thirteen vertebrae, but a single one could be 25 centimeters long! This animal's very strange anatomy took paleontologists a long time to reconstruct correctly, and they still cannot decide what kind of lifestyle it might have followed. Some see it as a terrestrial animal, while others argue that with such a neck it could have been at home only in the water. Study of the articulation of *Tanystropheus*'s vertebrae has moved scientists to argue for a lifestyle that was mostly aquatic; its extraordinarily long neck would have been incapable of elevation much beyond the horizontal and therefore would have made it very difficult for the animal to move about on land. However, *Tanystropheus* had no obvious adaptations to aquatic life. Its tail was flattened top and bottom, not laterally, and was enlarged at the base; such a configuration is not found in animals that use an undulatory swimming style. Moreover, its forelimbs were much shorter than its hind limbs, an arrangement that is compatible with neither paddling nor rowing. *Tanystropheus* might have inhabited the shoreline, then, moving from rock to rock in search of food, stopping regularly to plunge its long neck into the water and emerging with a mouth full of fish or shellfish. Examination of a specimen in which some soft tissue has been preserved seems to suggest a very fleshy tail. This extra mass would have helped balance the weight of *Tanystropheus*'s enormous neck, even when it was out of the water.

Macrocnemus, a small archosauromorph with a slender body (fig. 4.4), is known from only two species and, to this day, has been unearthed only at Monte San Giorgio and in China (Yunnan). This animal, although no more than 80 centimeters long, had a long neck as well. Just as in the case of *Tanystropheus*, its lifestyle is not clearly established: some experts think it lived on land, where probably it chased insects; according to others, because its remains are found only in sediments from coastal environments, it must have been adapted to an aquatic environment.

► Fig. 4.7. The Upper Triassic deposit in Guanling, Guizhou province (China).

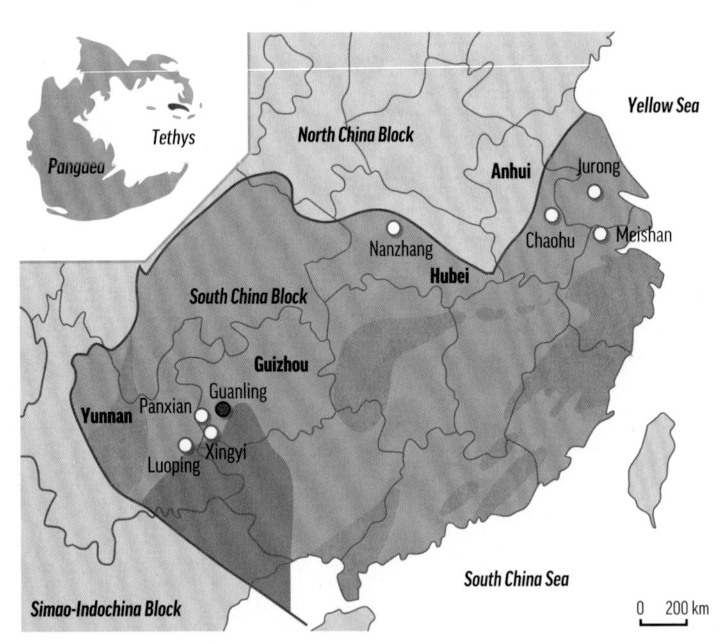

▲ Fig. 4.8. Part of the Middle Triassic Tethys, enclosed in a red border, showing areas of dry land (brown), shallow seas (medium gray), and deep seas (dark gray), atop the corresponding area of southeast China. The principal deposits containing marine reptiles (Guanling included) are in the provinces of Yunnan and Guizhou in the west and Hubei and Anhui in the east.

Guanling

A Profusion of Sea Lilies

Located in China's southwest, in the province of Guizhou (fig. 4.8), the Nanpanjiang sedimentary basin is largely represented by geological formations around 247 to 220 million years old, deposited from the end of the Early Triassic to the beginning of the Late Triassic. It is in one of these geological formations, the Xiaowa formation, that Guanling's fauna was discovered (fig. 4.7), a unique **Lagerstätte** (a sedimentary deposit very rich in fossils) of marine reptiles and crinoids.

Deposited around 220 million years ago (in the Carnian age), Guanling's marls contain an incredible association of vertebrates and invertebrates. Discovered at the beginning of the 1940s, the Guanling deposits first garnered attention thanks to the spectacular fossil crinoids, also called sea lilies, which can also be found (though represented by different species) in another deposit that is exceptionally well preserved—namely, Holzmaden (chapter 5, p. 138). These echinoderms, closely related to sea urchins, are equipped with large, flexible arms that allow them to filter the plankton they feed on from the water. Their lifestyles are fairly varied; some are mobile, while others are fixed to a support by means of a stem that varies in length. Guanling's sea lilies seem to have decided to both be mobile and remain fixed: several of the species in these deposits attached themselves to pieces of wood, which probably floated on the surface of the water.

The farmers who fortuitously discovered these crinoid fossils sold them on the black market. Yet it was the discovery of marine reptiles that brought wide attention to these geological layers. Because the commercial value of these vertebrate fossils was much higher than that of the sea lilies, exploitation of these deposits exploded and the discovery of colonies of crinoids and of complete, articulated skeletons of marine reptiles multiplied. In certain places the deposits are so rich in fossils that every hundred square meters contains at least one or two marine reptiles and three or four colonies of crinoids! Unfortunately, the exploitation of this very prolific deposit was such that the local environment was pillaged, the excavations leaving nothing but sterile rock. Thankfully, since 2003, a wide protective area has been established, and the formation is now part of a geopark, allowing for the restoration of the environment around the fossil-bearing deposits. There is a museum on site, where visitors can admire numerous fossils collected from nearby.

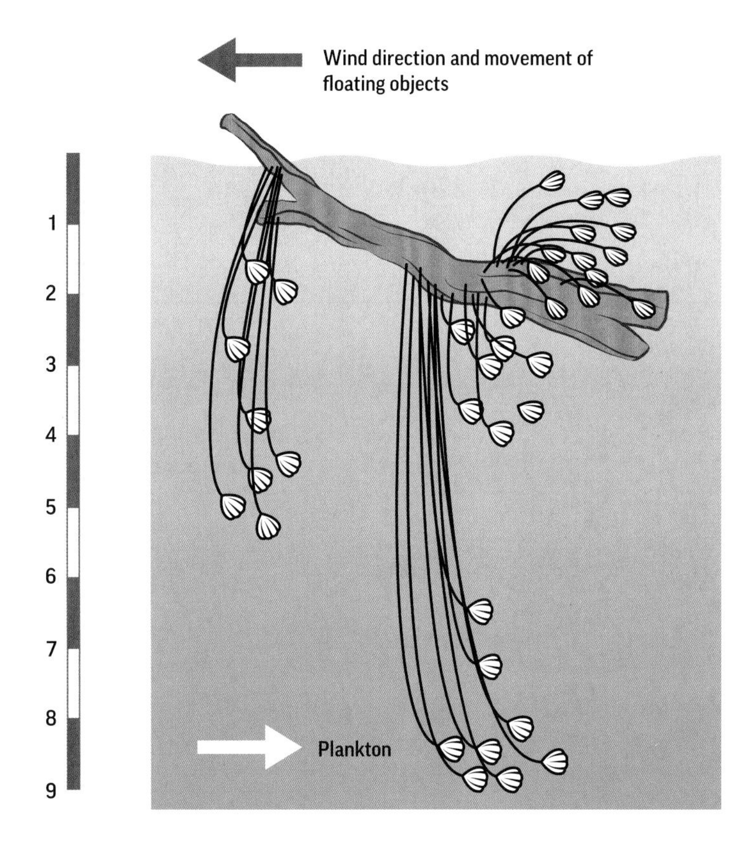

▲Fig. 4.9. Some crinoids attached themselves to driftwood. These animals were thus both sessile and mobile!

▼ Fig. 4.10. (Following pages) A Late Triassic seascape, Guanling, Guizhou province (China). Two ichthyosaurs, *Guanlingsaurus* (in the center) and *Qianichthyosaurus* (on the right), as well as the thalattosaur *Miodontosaurus* (on the left), swim between colonies of *Traumatocrinus* crinoids. *Psephochelys*, a small placodont (bottom right), and two actinopterygian fish, *Birgeria* (top right), complete the scene.

The Open Ocean as Horizon

The Xiaowa formation covers more than 200 square kilometers, in an area that was once a large inlet of the Tethys. The sediments were deposited in a calm marine environment, one that was stagnant and deep, during a period of maximum transgression (rise in sea level), which helps explain why the fossils are almost exclusively of pelagic (open-ocean) animals. The stillness of the water resulted in its being poorly oxygenated at the depths to which animals' remains settled, which helps explain the excellent condition of the fossils—in a low-oxygen environment, decomposition is very slow. Plus, because oxygen was scarce, only a few specialized bivalves—no scavengers—lived here.

Near the surface, however, both invertebrates and vertebrates abounded. The absence of wood-boring forms of marine life meant that dead wood might float along the currents for years. Colonies of *Traumatocrinus* crinoids lived attached to these pieces of

Top Predator: *Shastasaurus*

Name: "lizard from Mount Shasta" (Siskiyou County, California, United States)
Classification: Ichthyopterygia, Ichthyosauria, Merriamosauriformes, Shastasauridae
Lived: Late Triassic (Carnian–Norian, 229–204 Ma)
Known range: southeastern China, western United States (California), western Canada (British Columbia)
Overall length: up to 21 meters
Diet: fish, cephalopods

Shastasaurus was not only the largest ichthyosaur, but the largest marine reptile ever. Four species are known: *S. pacificus* is the earliest large ichthyosaur of which bones still connected at the joints (i.e., articulated bones) have been found. The giant *S. sikanniensis* was previously attributed to the genus *Shonisaurus*, which also includes a number of large species (up to 15 meters long). *S. liangae* exhibits the most vertebrae (86 pre-sacral and 110 caudal) of any ichthyosaur and one of the highest numbers among amniotes. The status of *S. neoscapularis* is very much debated because the few specimens that exist do not establish its anatomy very well.

All *Shastasaurus* exhibit a long, relatively slender body and a long tail with an upper lobe that was probably much less developed than in later species, and all were enormous. Except for some juvenile specimens, they possessed an unusually short and toothless snout, as well as a very thin lower jaw. Given those specializations, they must have fed using suction, in the manner of today's beaked whales.

► Fig. 4.11. *Shastasaurus*, an ichthyosaur from the Upper Triassic in North America and China.

wood (fig. 4.9): some crinoids could grow to gigantic dimensions, with stems 11 meters in length! Schools of fish and ammonites could find protection and nourishment among these colonies, which provided the nutritional foundation for one of the greatest known concentrations of marine reptiles from the Upper Triassic.

Following the Crisis, Sunny Days!

Dozens of specimens of marine reptiles have been found in the marls of Guanling, including at least three species of ichthyosaurs, five species of thalattosaurs, two species of placodonts, and one turtle. The latter, named *Odontochelys*, is utterly exceptional from an anatomical point of view (it had teeth and a plastron but no carapace) and provides several insights into the evolution of the first turtles (see chapter 2, p. 75). The placodont *Psephochelys* was also equipped with a shell and, like other placodonts, with flat teeth for crushing prey. The ichthyosaurs were, for their part, very different from one another, ranging from forms less than 2 meters long, such as *Qianichthyosaurus*, to the 10-meter giant *Guanlingsaurus*. The latter had a particularly small head, less than 10% of the animal's total length, as well as toothless jaws, which suggest that it was a suction feeder. The thalattosaurs were an extremely diverse bunch—*Miodontosaurus* must have exceeded 4 meters in length, and *Xinpusaurus*, about 2 meters in length, had a strange, pointed rostrum. The great variation in size among these reptiles, from the 40-centimeter tortoise to an ichthyosaur that exceeded 10 meters in length, and the variety of their apparent feeding habits, reveal that they filled a wide spectrum of ecological niches.

The faunas discovered in the Nanpanjiang basin extend from the end of the Early Triassic to the beginning of the Late Triassic, while transitioning through the beginning and the end of the Middle Triassic—in other words, we find four different faunas all together. With its wealth of fossils and because it encompasses most of the Triassic, this succession is an incredible tool for evaluating the changes in fauna and the recovery of ecosystems following the great Permian/Triassic crisis, as well as the radiation of certain groups, such as the marine reptiles, and it allows us to observe the great increase in their variety as well as their size over the course of the Triassic. After reptiles' massive invasion of the seas roughly 4 million years after the crisis, the recovery of the ecosystems seems to have taken at least 20 million years. These studies must, of course, be evaluated with caution, because phenomena observed in one basin in China cannot necessarily be extrapolated to a global scale: numerous local factors can affect these biotopes, independently of global factors. Nonetheless, it is by multiplying these kinds of finds and comparing the evolution of environments around the globe that we will be able to draw conclusions about the modalities and pace of the recovery of environments that were subjected to significant ecological stress. In addition, this data can provide us with important information on later crises—for instance, the current mass extinction.

The Secrets of Bone: A Window into the Biology of Extinct Animals

Studying the interior of bones allows us to access certain kinds of information about the biology of once-living creatures. It can be carried out in two distinct fashions: the microanatomical approach looks at the structure and distribution of the bony tissue, to help us understand animals' ecology, while the histological approach looks at the nature of the bony tissue, to enlighten us regarding their metabolism.

Microanatomy

Bones ensure mechanical support, protect vital organs, house the spinal cord, provide sites for muscles to attach to, and serve as the principal calcium reservoirs in the body. Bones' rigidity is not directly tied to their mass; instead it depends on the organization of the bony tissue. Contrary to popular belief, bone is a dynamic tissue that is subject to modification (i.e., it undergoes phases of resorption and deposition) and constant repair over the course of the animal's life. This allows bone not only to adjust and maintain its shape while it grows (fig. 4.12), but also to maintain an appropriate balance of strength, so as not to fracture, and lightness, to favor mobility. The internal structure of bone (its microanatomy) reflects the biomechanical constraints it is subjected to and therefore relates to the animal's lifestyle.

During fossilization, the internal structure of bones is generally preserved, and preserved so well, that this information is just as available from fossilized bones as it is from organic bones. The study of the internal structure of bone in today's vertebrates has allowed us to underscore the ties between the microanatomical characteristics of bone and the ecology of the animal. It therefore can help us infer the lifestyles of fossil organisms.

In animals of terrestrial descent that have adapted to the aquatic environment, we observe one of two types of bone specialization: an increase in bone mass or the development of a very spongy internal structure.

The first, which is characterized by an inhibition of the processes of bone modification and/or by an increase in bony deposits (at the bone's periphery or at its core in the course of modification), ensures passive hydrostatic (i.e., without the need for swimming movements) control of buoyancy and pitch angle (see "Adaptations to the Aquatic Environment," pp. 118–19) in relatively inactive swimmers (e.g., sirenians, pachypleurosaurs, and mesosaurs). The increase in bone density is called **osteosclerosis** (fig. 4.13, center), and the addition of bony deposits at the bone's periphery, which gives the bone a bloated appearance (fig. 4.14), is pachyostosis. The combination of the two is called pachyosteosclerosis.

The second response involves the transformation of the entire internal structure to a spongy texture (fig. 4.13, right)—in other words, the bone appears as almost entirely composed of bony plates. The network of bony plates is,

▼ Fig. 4.12. 1. During growth, the process of bone modeling preserves the shape of the bone. 2. Over the course of an animal's life, the process of remodeling helps it adapt to functional constraints. 3. The mechanism of bone remodeling involves several steps.

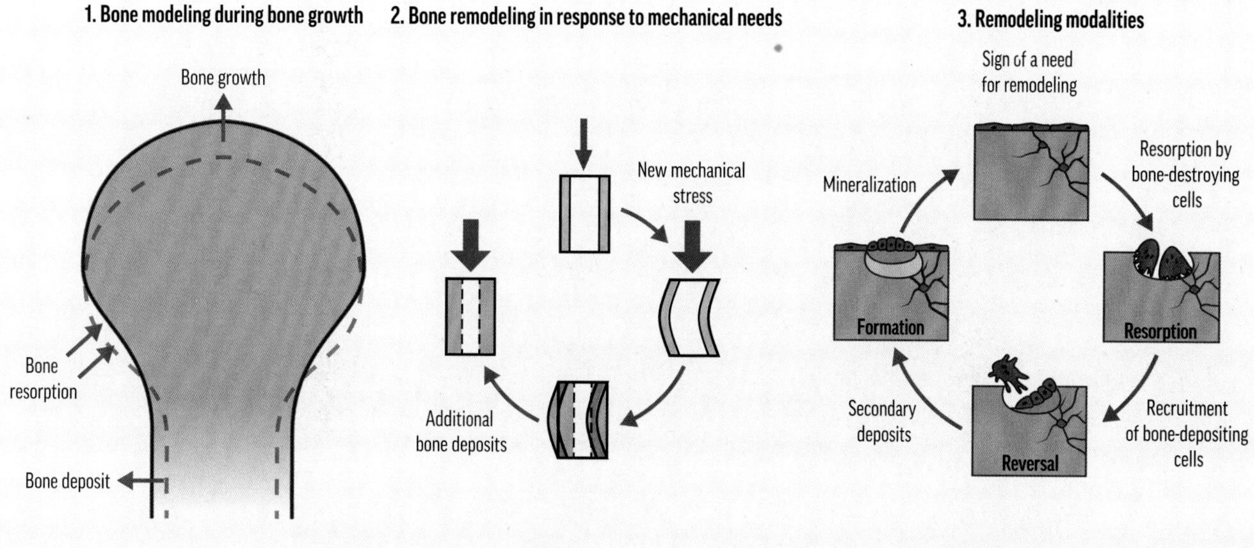

1. Bone modeling during bone growth
Bone growth
Bone resorption
Bone deposit

2. Bone remodeling in response to mechanical needs
New mechanical stress
Additional bone deposits

3. Remodeling modalities
Sign of a need for remodeling
Resorption by bone-destroying cells
Resorption
Recruitment of bone-depositing cells
Reversal
Secondary deposits
Formation
Mineralization

◄ Fig. 4.13. Cross sections of vertebrae (above) and bones (below) exhibiting osteosclerotic textures (center) and spongy textures (right) as compared to more tubular structures (left) observed in terrestrial reptiles. Taken from the following creatures (top, left to right): a monitor lizard, a basal ophidiomorph, an ichthyosaur; (bottom, left to right): *Dallasaurus* (a mosasauroid), an unknown eosauropterygian, *Plotosaurus* (a mosasauroid).

▼ Fig. 4.14. Different views of two dorsal vertebrae of a basal ophidiomorph ("dolichosaur"). The increase in bone mass of the vertebra at bottom, by means of the addition of peripheral bone deposits (pachyostosis), gives it a bloated appearance.

in this case, fairly consistent: there are no large cavities (which would constitute significant concentrations of constraints, nor any **medullary** cavities in long bones. In this fashion, the stresses placed on the bone during locomotion are more evenly divided, which increases its resistance to breaking. This resistance is also generally improved if the bony plates are oriented in the direction of maximal force exerted during swimming. This specialization seems to be ideal in an aquatic environment.

In the case of animals with bodies that are poorly adapted to an active and rapid swimming style, control of buoyancy and of bodily angle cannot be accomplished hydrodynamically (i.e., by making swimming movements), because this would be too costly in terms of energy. These creatures therefore need to achieve an increase in bone mass that will facilitate their relatively slow mode of swimming for long stretches at shallow depths, even if this specialization limits their movement abilities. For active swimmers capable of controlling their buoyancy and bodily angle in a dynamic fashion (such as cetaceans, mosasaurs, and plesiosaurs), however, an increase in bone mass is no longer necessary and, moreover, would create a real handicap by decreasing maneuverability and acceleration. A spongy bone structure, instead, provides their skeletons with the lightness and the strength these creatures need.

Histology

Bone is a connective tissue formed by an organic matrix essentially composed of collagen fibers that are progressively mineralized into calcium phosphate. The study of the characteristics of tissue (histology) of bones provides a wealth of biological information.

The nature of bony tissues is determined, first of all, by the organization of the collagen fibers; by the importance of the characteristics of the vascular system (blood vessels) (i.e., the nature and orientation of the vascular

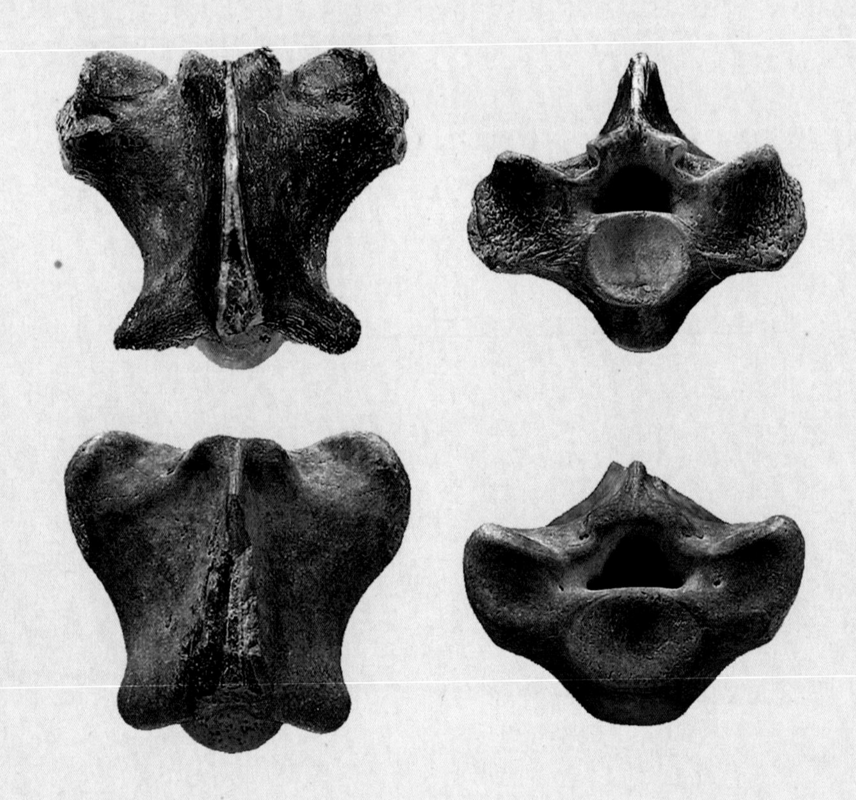

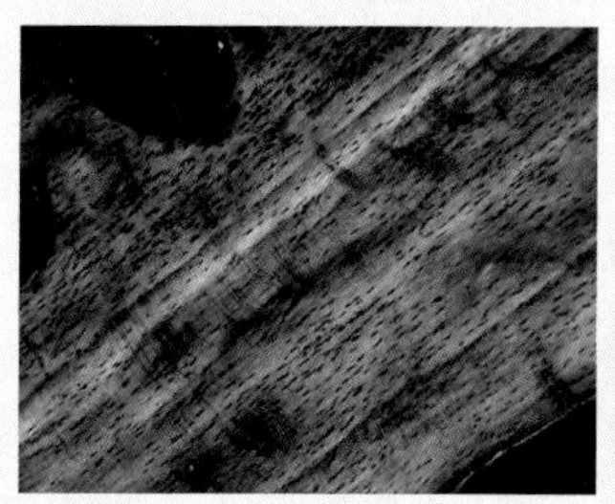

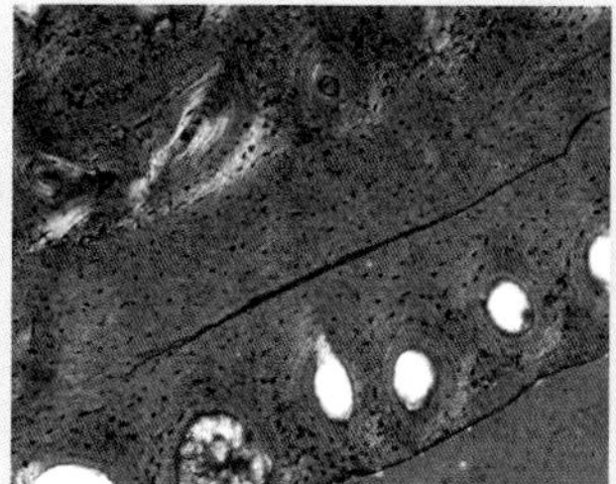

▲ Fig. 4.15. Bone tissue from different animals showing different types of vascularization: 1. Non-vascularized tissue (in monitor lizards). 2. Simple vascular channels with a radial orientation (in anacondas) as well as some anastomoses (connections between channels). 3. Primary osteons (simple vascular channels with centripetal deposits of bone tissue) oriented longitudinally (in mosasaurs). 4. Longitudinally oriented osteons with numerous anastomoses (connections) that are essentially circumferential (in ichthyosaurs).

channels within the bones); and by the size, shape, and orientation of the bone cells (fig. 4.15), all of which provide information as to the bone's speed of growth. For example, a bone's rate of growth generally increases with the degree of vascularization: the more blood vessels, the faster the growth. And, conversely, growth rate decreases as a function of the bone's degree of organization—in effect, the longer it takes for bone deposition to occur, the more organized it can be. Therefore, collagen fibers that are well organized with respect to one another, arranged at specific angles (lamellar tissue), and bone cells that are oriented in the direction of deposition, indicate a slow deposition process. Conversely, if the deposition is rapid, the collagen fibers and the bone cells are oriented in random fashion (fibrous tissue).

Lamellar tissue is generally found in modern reptiles (except birds), whereas fibrous tissue can be found in animals that exhibit rapid growth, such as birds and mammals but also dinosaurs, pterosaurs, and ichthyosaurs. The rate of growth is deemed to be partially tied to the animals' basal metabolism. It is therefore possible, using histological data, to make some inferences about the metabolism of fossil vertebrates. This must be done cautiously, because the connection between histological data and metabolism is indirect. This approach has, notably, been used to suggest that some dinosaurs were "warm-blooded."

Bone, just like wood, has a speed of deposition that evolves cyclically, to the extent that signs of growth similar to the rings of a tree trunk can be observed (fig. 4.16). These markers consist of a combination of (1) zones, wide layers of rapid growth (during periods when resources are most abundant); (2) annuli (the plural of annulus, meaning "ring"), narrower layers that indicate a slower rate of growth; and (3) lines, indicating periods of no growth. Since a new marker is created each year, examination of these markers allows us to determine the age of the specimen being studied and therefore the age of the creature when it died. It can be very useful to know, for example, whether we are looking at a juvenile or an adult. Examination of a sufficient variety of specimens of the same species even makes it possible to determine the age at which members of that species attained sexual maturity.

The histology of fossilized bones (paleohistology) therefore provides valuable biological data that anatomical study of the skeleton alone generally cannot.

▼ Fig. 4.16. Lines showing arrested growth (indicated by arrows and dotted lines) in the rib (left) of a pachypleurosaur, *Neusticosaurus*, and the vertebrae (right) of an unknown eosauropterygian (above) and of an ophidiomorph ("dolichosaur"), *Coniasaurus* (below).

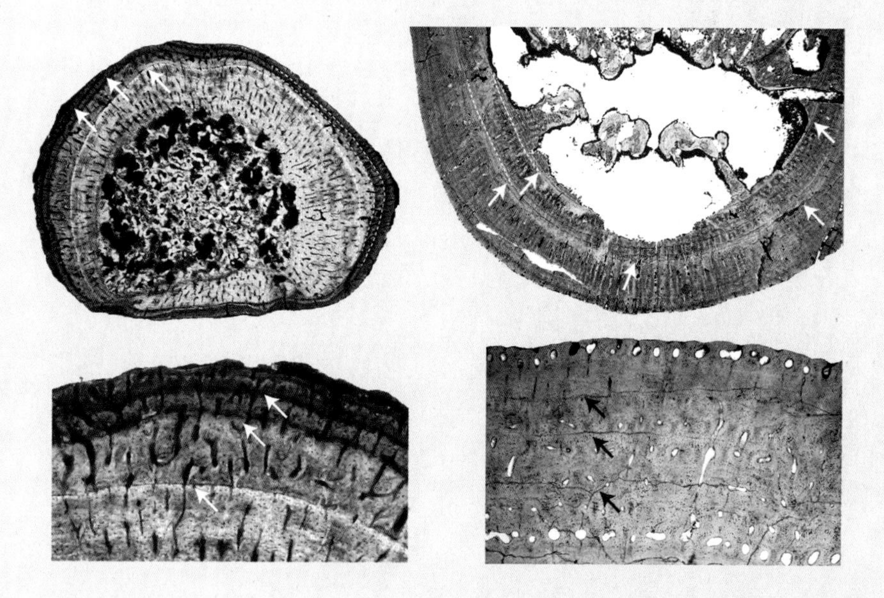

Atopodentatus

Atopodentatus unicus, nicknamed "the lawn mower of the Triassic," was a Middle Triassic marine reptile with a very unusual arrangement of teeth, from which its name, loosely meaning "bizarre, unique dentition," derives. The formation in which it was discovered, in Yunnan province, China, has provided several rich faunas of marine reptiles from across the Triassic, including ichthyosaurs, several sauropterygians (pachypleurosaurs, nothosaurs, and placodonts), some archosauriforms, and finally some saurosphargids, a group of diapsid reptiles with unclear affinities.

Based on genetic analysis, *Atopodentatus* was a sauropterygian of some kind, but paleontologists are unable to narrow it down. The species' description, published in 2014, is based on a nearly complete skeleton, with a laterally flattened and very poorly preserved skull. In 2016 two new specimens, with very well-preserved skulls, were described. They perfectly illustrate how differences in preservation can influence, and even radically alter, our reconstructions of extinct animals (fig. 4.17).

Atopodentatus lived around 245 million years ago, in the Anisian age. It was almost 3 meters long, with a long body, a very small head, and legs rather than swimming paddles. It exhibits a solid sacral-pelvic connection and a weak ossification of the wrists and ankles, indicating that it was aquatic but occasionally moved about on dry land. Its phalanges were shaped like little hoofs, resembling those of the placodont *Psephoderma*, a really rare occurrence among the marine reptiles of the Mesozoic!

Independently of the interpretation that has been proposed for its snout (fig. 4.17), *Atopodentatus* was characterized by a surprising dentition, with more than 175 teeth on the upper jaw and 190 on the lower. These teeth were arranged in rows, like those of hadrosaur dinosaurs. Very fine and slender, they call to mind the teeth of mesosaurids and certain pterosaurs (e.g., *Ctenochasma* and *Pterodaustro*), except that *Atopodentatus*'s teeth were more bladelike at the base and more needlelike at the tip, and they exhibit some partial overlap as well.

The first specimen to be described left paleontologists perplexed. The animal was, in effect, equipped with an astonishing vertical rostrum loaded with teeth that gave it the appearance of a zipper, in addition to a massive mandible with a shovel-like end. This anatomy, which was strange to say the least, led experts to conclude that *Atopodentatus* had probably been a burrowing animal that looked for food on the seafloor and could resort to a beak, a shovel, and some hooves ... like a fanciful cross between a flamingo and a hippopotamus.

The second and third specimens exhibited a snout that was at least as strange, but, since it had not been deformed by a significant lateral flattening, shed new light on its shape and function. The snout was in fact not vertical but horizontal; was very short and extended at the sides, as in hammerhead sharks; and resembled the quadrangular snout of the placodont *Henodus* (figs. 2.21 and 2.22, p. 44). It contained very fine teeth that could cut like scissors, described below, which allowed it to graze on the algae of shallow seabeds. The teeth in the rear part of the jaws, arranged like a tight trap, probably functioned as a filter. Just like *Henodus*, however, *Atopodentatus* does not exhibit any of the characteristics of true filter feeders. Its very particular oral apparatus does somewhat resemble that of the sauropod dinosaur *Nigersaurus*, a little wink from Mother Nature to draw our attention to a way of feeding that must have been very close to that of our aquatic "lawn mower"!

Of the hundreds of varieties of Mesozoic marine reptiles, only two seem to have been vegetarian, and they both come to us from the Triassic: the placodont *Henodus* (from the Carnian) and *Atopodentatus*. Next to its many carnivorous contemporaries, large and small, which occupied many ecological niches, *Atopodentatus* represents not only the most ancient vegetarian marine reptile but also evidence that the food chains were very rapidly reestablished after the Permian/Triassic crisis.

▼ Fig. 4.17. Different reconstructions of *Atopodentatus*'s snout, as a function of the ways in which the specimens had been preserved. On the left: the "zipper" model, based on a skull found in 2014, and the comparable ecological model, a flamingo (a filter feeder). On the right: a "lawn mower" model, based on a skull found in 2016, and a comparable ecological model, the grazing sauropod dinosaur *Nigersaurus*.

An Underrated Crisis

In the annals of biological crises, the one that unfolded at the end of the Triassic period 201 million years ago remains, for obscure reasons, largely unrecognized. And yet it was similar in magnitude to the Cretaceous/Paleogene crisis; many terrestrial species suffered extinction as a consequence of it. Although dinosaurs, crocodylomorphs, and pterosaurs emerged relatively unscathed, such was not the case for archosauromorphs: the large terrestrial predators, such as the rauisuchians; freshwater piscivores, such as the phytosaurs; and the herbivorous aetosaurs (see figs. 1.3 and 1.4, pp. 10–12) did not survive. The same fate awaited the last therapsids (once called "mammalian reptiles"), dicynodonts, and cynodonts, already significantly affected by the Permian/Triassic crisis; most were extinguished at the end of the Triassic, and their only legacy as far as descendants are concerned was the lineage that would give rise to mammals, which existed mostly beneath notice until the Cenozoic era. And the giant amphibians—metoposaurids, mastodonsaurids, and capitosaurids—didn't stand a chance. These extinctions on land would allow the dinosaurs to assert their supremacy for the remainder of the Mesozoic.

In the oceans, also, times were tough: 76% of the species—including all conodonts, "primitive" vertebrates that had arisen in the Precambrian era; several entire families of ammonites; and some bivalves—disappeared, and coral reefs were significantly affected too. As far as marine reptiles are concerned, while the placodonts actually did go extinct at the very end of the Triassic, it was mostly in the intervening time between the last two ages of the Triassic, the Carnian and the Norian, or 228 million years ago, that the great biotic upheavals took place: we do not see evidence of sudden extinction, but we see major faunistic renewals over the course of the Late Triassic. The pachypleurosaurs and the nothosaurs, already reduced in number of species, went extinct, while other groups branched out or diversified. The plesiosaurs began to impose themselves in the seas, while the ichthyosaurs continued their conquest of the open oceans. Somewhat later, in the Early Jurassic, the thalattosuchian crocodyliforms would appear.

The causes of the crisis are hard to ascertain, but there are two popular hypotheses. The one that is advanced most often has to do with the breakup of the supercontinent Pangaea and the formation of what would become the Atlantic Ocean. The new ocean, the Central Atlantic, needed filling up with water, which would have lowered the global sea level. The resulting shrinkage of the epicontinental seas would have caused the extinction of marine reptiles that were mostly coastal, such as placodonts and the last nothosaurs, due to habitat reduction. Conversely, it would have favored the branching of pelagic reptiles, such as the parvipelvic ichthyosaurs (those that flourished during the Jurassic and Cretaceous) and the plesiosaurs. In addition, the fracturing of a continent involves intense vulcanism at the site of the rupture, raising magma that will cool to form oceanic crust. As the Central Atlantic opened, a gigantic magmatic area of around 7 million square kilometers (nearly the size of the contiguous United States), called the Central Atlantic magmatic province, or CAMP, formed. The eruption of more than 2 million cubic kilometers of basalt, comparable to the volume of the Deccan Traps from the Cretaceous/Paleogene boundary, would have created an array of ecological threats.

In the short term, the release of sulfur dioxide (SO_2) and other atmospheric pollutants would have acidified the oceans, thereby affecting biocalcification (the formation of shells and skeletons) in sea creatures, and globally cooled the climate by blocking sunlight. Once more, there would have been an opposing trend too—namely, the massive release of carbon dioxide (CO_2) would have led, in the long term, to a very significant global warming, destabilizing ecosystems even further.

The other hypothesis points the finger at asteroid impact. Several craters have been dated to between 214 and 219 million years ago, or shortly before the Triassic/Jurassic crisis: Manicouagan crater (100 kilometers wide), in Quebec (Canada); Saint Martin crater (40 kilometers wide), in Manitoba (Canada); and Rochechouart crater (about 50 kilometers wide), near Limoges, in France. The latter was previously dated to around 214 million years ago—therefore too old to be responsible for the crisis—but recent research has rejuvenated it by 12 million years! It is therefore a good candidate. Its size, which is modest compared to the 180-kilometer-wide Chicxulub crater from the Cretaceous/Paleogene boundary in Mexico, does leave doubt about its alleged global impact. However, researchers noticed that, when placed on a map of the Triassic world, the craters of Manicouagan, Saint Martin, and Rochechouart are aligned, suggesting that the impacts were part of a single event in which an asteroid fractured, striking multiple places along its line of approach. The contemporaneous impact of three fragments, creating craters of 100, 40, and 50 kilometers in diameter, would have had much greater consequences than a single impact would! Yet so far, the craters don't appear to be of the same age.

But why look for a single cause? Many impacts by extraterrestrial objects in Earth's history did *not* lead to extinction events. The trigger of this biodiversity crisis may, once again, have been the conjunction of several factors.

CHAPTER 5

THE JURASSIC PERIOD

"L'ichthyosaure et le plésiosaure," illustration by Éduard Riou for *La Terre avant le Déluge* [Earth before the Flood], Louis Figuier, 1863

I

A New World

The Jurassic, the second period of the Mesozoic era, extended from 201 to 145 million years ago. Its name derives from the Jura, a mountain range in western Europe, where it was defined in 1829 by French geologist and naturalist Alexandre Brongniart. It is divided into three parts (Lower, Middle, and Upper), which initially were assigned the names—still in occasional use today—of Lias, Dogger, and Malm. Following the great faunistic renewal at the end of the Triassic (see chapter 3, p. 134), the thalattosaurs disappeared; the Triassic sauropterygians (nothosaurs, pachypleurosaurs, and placodonts) were replaced by the plesiosaurs, which had already made a timid appearance during the Late Triassic; and the non-parvipelvic ichthyosaurs also disappeared—the only ichthyosaurs that persisted were those adapted to the open ocean and capable of swimming long distances. The beginning of this period was characterized by the rise of another group of marine reptiles emblematic of the Jurassic, the thalattosuchian crocodylomorphs.

It was during the Jurassic that the well-known trio of marine reptiles—plesiosaurs, ichthyosaurs, and thalattosuchians—became established. These faunas, which were greatly different from those of the Triassic, lasted throughout the Jurassic. Even though the groups did not change in a large sense, their respective proportions did vary over the years, and by the end of the period the species in each group were quite distinct from those that had represented it at the beginning. We will discuss these subtle changes in the composition of these faunas of marine reptiles at the end of this chapter.

Holzmaden

Geologic Context

The Holzmaden region extends from the lower end of the Swabian Jura, a mountain chain that stretches more than 200 kilometers in Baden-Württemberg in Germany's southwest (fig. 5.2) and, together with the Franconian Jura (a mountain chain in Bavaria), forms a part of the entire Jura massif, which continues to the west in northern Switzerland and

◀ Fig. 5.1. Holzmaden deposit, Germany, 1912. Bernhard Hauff (1866–1950) and his team unearth a specimen of the ichthyosaur *Eurhinosaurus* in the family quarry.

in eastern France. The Swabian Jura is made up of sediments from the Jurassic period—this is where the name "Jurassic" comes from. It is divided geologically into the Lower Jurassic (Schwarzer Jura), the Middle Jurassic (Braune Jura), and the Upper Jurassic (Weisse Jura). The German names of these divisions refer to the dominant color of the sediments: black, brown, and white. The final stage of the Lower Jurassic, the Toarcian, is known the world over for the wealth of fossils to be found in this region. And we do indeed find numerous, extremely well-preserved fossils in these layers, which correspond to Konservat-Lagerstätten. This exceptional preservation is the result not of chance but of environmental conditions that prevailed during the Toarcian in this region.

Holzmaden, like most of Europe at that time, was covered by a shallow sea in which numerous marine reptiles proliferated. The marine sediments correspond to bituminous schists, called "schistes-cartons" in French, "jet-rock" in English, and "Posidonienschiefer" (Posidonia shale) in German. These are very fine-grained sedimentary rocks that vary in color from gray to black, are rich in organic matter, are thinly layered, and were formed in a very oxygen-poor environment. The somber color of bituminous schists is due to the presence of organic carbon, which corresponds to algae residues, bacteria, and other organisms that lived in the sea.

Very Particular Environmental Conditions

Marine environments in which oxygen is scarce are unfavorable for the survival of most organisms but splendid for the preservation of fossils (fig. 5.3). During the Toarcian (174–183 Ma), qualities of the sea in the Holzmaden region were quite different at different depths; we

▼ Fig. 5.2. A map showing the range of outcrops of Posidonia shale (areas shaded in black) in the southwest of the German basin. The deposits richest in vertebrate fossils are in and around the village of Holzmaden.

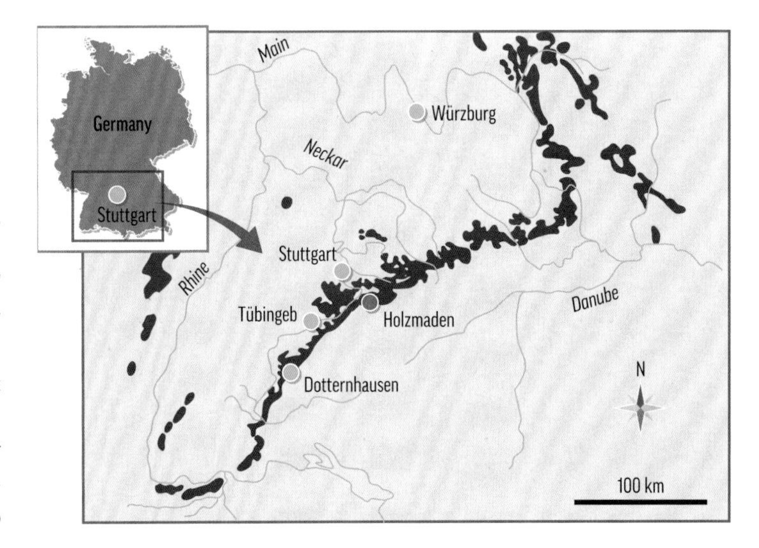

might say the sea was stratified, or distinctly layered, because the waters of each level did not mix with those above or below. The vast majority of organisms proliferated in the top layer, which, since it was in contact with air, was rich in oxygen. Phytoplankton, composed of microscopic algae (coccolithophores, dinoflagellates, and green algae), cyclically bloomed here, periods of massive proliferation alternating with those of just as massive die-off once all the nutrients had been consumed, thus dragging large amounts of organic remains toward the bottom. In the lower layer of water, the zooplankton and the bacteria that fed on those remains took their turn burgeoning too, consuming the little oxygen available. Since the significant stratification of the waters prevented oxygen (whether from the air or produced by algae) from reaching the lower layers, there was enduring oxygen deprivation in the deeper waters. In these conditions the degradation of the organic matter derived from the phytoplankton slowed very considerably, and it accumulated on the seabed, producing thick layers of black, waterlogged silts. The remains of larger organisms, such as ammonites and vertebrates, sank either partially or completely into this silt, where they degraded only very slightly. We can observe these kinds of environmental conditions today in the depths of the Black Sea or the Baltic.

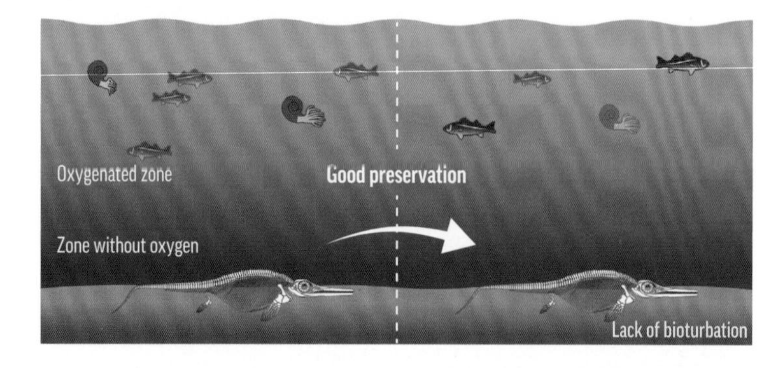

▲ Fig. 5.3. **Above**: In a marine environment, the sediments fall to the bottom and start covering the remains of animals. Even after rapid burial by sediments, the remains are subject to some degradation. The organic parts of the remains constitute some of the favorite food of many necrophages (macro- or microorganisms) and are therefore rapidly destroyed. The mineralized parts, such as bones and shells, are more difficult to break down and thus can sometimes be preserved either wholly or in part.
Below: Very scarcely oxygenated environments, although unfavorable for most forms of life, are ideal for the preservation of both mineralized and soft tissues. If, in addition, the remains are covered by a fine and impermeable sediment (e.g., silt), the chances that some traces of soft tissue will be preserved go up: the compact sediment keeps out oxygen and therefore provides extra protection against decomposition by microorganisms. The combination of these two factors resulted in the preservation of the contours and imprint of the skin of several ichthyosaurs in the Lower Jurassic deposits of the Holzmaden region (Germany).

The scarcity of oxygen near the seafloor of the Holzmaden region during the Toarcian led to the development of very specialized microorganisms: sulfate-reducing bacteria, which, by their metabolic activity, released hydrogen sulfide (H_2S). As it was being buried, this hydrogen sulfide combined with the iron in the detrital minerals to form pyrite (FeS_2), which today abounds either in the form of concretions or via the mineralization of soft tissues.

These anoxic conditions in the depths must have been interspersed with some short episodes in which the waters of different levels mixed (e.g., after a storm), leading to the temporary reoxygenation of the lowest waters. This would explain the numerous bivalve fossils, occasional sea urchin fossils, and tunnels created by burrowing organisms (e.g., crustaceans and worms) that have been found in several layers.

Exceptional Fossils

The incredible fossils of the Holzmaden region were discovered by chance, in the course of

▲ Fig. 5.4. A juvenile *Stenopterygius* (about 1 meter long) from the Lower Jurassic in Holzmaden, Baden-Württemberg (Germany), Muséum national d'Histoire naturelle (Paris, France). The trace of the skin, in the form of a thin black film around the bones, perfectly outlines the animal's contours in life, most crucially at the level of the dorsal fin and the upper part of the caudal fin, which were particularly fleshy.

quarrying (fig. 5.1) that began in the eighteenth century with the goal of extracting mineral oil from the bituminous schists and finally resulted in the exploitation of the Fleins, a flat layer of shale that was excavated for use in paving, oven stones, and, more recently, interior decoration. The unearthing of the first fossils therefore was completely fortuitous.

The most celebrated fossils from Holzmaden are the marine reptiles. But many other animals lived here too. Although lamellibranchs (a class of bivalves) were rare, one lamellibranch genus, previously attributed to *Posidonia*, is very abundant in certain layers, which is why the Toarcian deposits from this region are called Posidonia schists. These bivalves seem to have had a great tolerance for oxygen-poor waters.

The creatures that lived closer to the surface have been found in abundance. The crinoids (or sea lilies), such as *Seirocrinus subangularis*, with their long stalks, were remarkable (fig. 5.7). These echinoderms (as well as various lamellibranchs) attached themselves to driftwood. When the wood eventually sank, it took the organisms with it down to the muddy seafloor. The most noteworthy colony of crinoids found was anchored to a 12-meter-long tree trunk! The fossils most commonly found in the deposit are those of cephalopods (ammonites and belemnites), represented by a very large number of species. Belemnites, which lacked an outer shell, had an internal "bone," as cuttlefish do. Called a rostrum, this bullet-shaped structure, together with the phragmocone (the thin internal shell), constituted the animal's skeleton. In other places, most belemnite fossils consist only of the rostrum; soft parts are rarely preserved. Yet in Holzmaden some complete specimens in which the animal's musculature, ink sac, and tentacles can all be seen have been unearthed. In addition, many fossils of cartilaginous and ray-finned fishes have been collected, the largest ones several meters in length—such as those of the shark *Hybodus*, some of which are so spectacularly preserved that they allow detailed anatomical study.

Holzmaden's fossils of marine reptiles (ichthyosaurs, plesiosaurs, and thalattosuchian crocodylomorphs) are particularly impressive. They are well preserved, often complete, with the elements of the skeleton generally connected to one another. The plesiosaurs are the least numerous of the reptiles found in these deposits—fewer than twenty specimens have turned up—and are of modest dimensions, not exceeding 4 meters in length. They are represented by five genera: the pliosauroids *Hauffiosaurus* and *Meyerasaurus* and

▲ Fig. 5.5. An extraordinary snapshot of Early Jurassic life, but an exceptionally cruel fate: a pregnant ichthyosaur and its embryo, expelled at the moment of death, fossilized in Holzmaden's exceptional deposit (Staatliches Museum für Naturkunde, Stuttgart, Germany) Stuttgart, Allemagne).

the plesiosauroids *Hydrorion*, *Seeleyosaurus*, and *Plesiopterys*. Holzmaden has yielded three genera of thalattosuchian crocodylomorphs: *Macrospondylus*, *Pelagosaurus* (fig. 2.48, p. 68), and *Platysuchus*. They represent about 15% of the reptiles found in this deposit. All three forms had a long rostrum, in the manner of today's gavials, an indication of a diet of fish. *Macrospondylus*, the largest of the trio, could grow to 7 meters long. The ichthyosaurs are without a doubt the most abundant reptiles in this deposit. More than three hundred magnificent specimens have been found, and many of them are on exhibit in museums around the world. They represent five genera: *Stenopterygius*, *Temnodontosaurus* (fig. 5.8), *Eurhinosaurus*, *Suevoleviathan*, and *Hauffiopteryx*. The largest *Temnodontosaurus* from this location were about 10 meters long!

In Flesh and Bones

In 1894, Bernard Hauff, the founder of Holzmaden's celebrated museum (the Urweltmuseum Hauff), after very detailed work on a perfectly preserved ichthyosaur specimen, exhumed the animal's contours, conserved in the form of an organic black film. He highlighted the presence of a dorsal fin and a bilobed caudal fin (fig. 5.4). The dorsal fin and the upper lobe of the caudal fin were not supported by any bone, and their presence had not been surmised prior to this discovery; ichthyosaurs had been represented in illustrations as lacking a dorsal fin and with a long, slender tail.

The preservation of soft tissue allowed researchers to deduce aspects of ichthyosaurs' biomechanics, physiology, and ecology as well. By meticulously studying the fossilized soft tissues surrounding the bones, they found that, in life, these animals had been almost entirely covered in a complex system of fibers. The fibers may have worked to keep their skin taut as they swam, preventing folds from forming in the skin when it flexed, in order to eliminate the small amount of drag that creates (see also "Locomotion in Ichthyosaurs and Sauropterygians," pp. 148–49). From a hydrodynamic point of view, this would have been an enormous advantage, allowing even faster swimming with less waste of energy.

Unfortunately, we have very few crocodylomorph fossils in which soft tissue has been preserved and only two such specimens of plesiosaurs from Holzmaden. This rarity is probably owing to the fact that these animals have not been unearthed in the colossal numbers that ichthyosaurs have.

▲ Fig. 5.6. The way in which ichthyosaurs were born: "headfirst" in the case of *Chaohusaurus* (from the Lower Triassic in China) versus "tail first" in the case of *Stenopterygius* (from the Lower Jurassic in Europe).

Females with Their Young

The large quantities of fossils and the presence of soft tissues in some of them made the site valuable to paleontologists, but what the Holzmaden deposits are truly famous for is skeletons of young ichthyosaurs in the abdominal region of some adult specimens. When they were first discovered, it was very quickly concluded that, owing to their completeness and the absence of signs of predation, these were embryos and thus these marine reptiles gave birth to live young. More than 150 embryos from the deposits of this region, all belonging to *Stenopterygius* (fig. 5.5), have been cataloged.

Signs of multiple impregnations are frequent, and up to eleven embryos have been found in the body of a single female! The smallest embryos have been found rolled up on themselves, whereas the more mature ones are extended, with their head toward the mother's front. The latter indicates that ichthyosaurs were born "tail first," as today's cetaceans are, probably to minimize the risk of drowning during farrowing (fig. 5.6).

Holzmaden's *Stenopterygius*, which were slightly over 2 meters long when fully grown, averaged 75 centimeters long at birth—in other words, about a third the length of the mother. Some *Stenopterygius* mothers have been preserved with an embryo that has only partially emerged from their body (fig. 5.5). It has been supposed that these individuals died while giving birth. However, it may be that they died late in pregnancy and the embryo was preserved in the process not of being born but of being expelled by the gases produced by putrefaction—this is a less moving history but the one that currently has more support.

▲ Fig. 5.7. (Previous pages). A Toarcian (183–174 Ma) Early Jurassic seascape, Holzmaden, Baden-Württemberg (Germany). On the right, *Temnodontosaurus*, an immense ichthyosaur, has just caught a small plesiosaur, *Seeleyosaurus*. On the left, a crocodylomorph, *Macrospondylus* (above), chases a school of teleostean *Leptolepis*, while *Stenopterygius*, an ichthyosaur, prefers some cephalopods. Near the surface, some *Seirocrinus*, sea lilies, drift attached to pieces of wood.

Ichthyosaurs exhibit very advanced adaptations to aquatic life. Given their large size and their overall anatomy, such as their short swimming paddles, if they had needed to lay their eggs on land, they probably would have become stranded like many of today's whales. If they had needed to lay their eggs in the water, that would have been problematic because the necessary exchange of gases (carbon dioxide and oxygen) through the semi-permeable eggshell is considerably slower in water than it is in the air (some rare turtles in the Chelidae family lay eggs underwater, at the muddy bottom of seasonal ponds, but the eggs don't develop—a phenomenon known as **diapause**—before the dry season, when the ponds disappear). Ichthyosaur embryos therefore developed in the womb. Such viviparity is known in about 20% of modern reptiles (if we exclude birds), as well as in many sharks, but also in other extinct marine reptiles: mosasaurs (see chapter 2, p. 90) and sauropterygians (chapter 2, p. 43).

Top Predator: *Temnodontosaurus*

Name: "lizard with cutting teeth"
Classification: Ichthyopterygia, Ichthyosauria, Merriamosauriformes, Temnodontosauridae
Lived: Early Jurassic (Hettangian–Toarcian, 200–176 Ma)
Known range: England, France, and Germany
Skull length: 2–3 meters
Overall length: 6–12 meters
Diet: cephalopods, fish, marine reptiles

This ichthyosaur is regarded as one of the Early Jurassic's top marine mega-predators. Preserved stomach contents point to a diet almost exclusively of vertebrates, including other marine reptiles. The first ichthyosaur Mary Anning collected, in 1811, was a nearly complete *Temnodontosaurus* (see "Mary Anning, One of the First Paleontologists," p. 40). One species, *T. azerguensis*, first described in detail in 2012, may have been toothless, which seems to suggest that some *Temnodontosaurus* did not follow the mega-predator diet and instead fed on prey they did not need to crush.

► Fig. 5.8. *Temnodontosaurus*, an ichthyosaur from the Lower Jurassic in Europe.

It therefore probably corresponds to a reproductive response to the environment's selective pressures and is tied to a high degree of specialization to aquatic life.

Nevertheless, the 2014 study of a fossil of a pregnant *Chaohusaurus* (one of the most ancient ichthyosaurs, from the Lower Triassic in China) with an embryo in a "head-first" position, as in terrestrial amniotes (fig. 5.6), shows us three things: (1) viviparity must have developed very early in the evolutionary history of the group, in the Triassic; (2) it must therefore have originated among terrestrial forms, rather than marine ones, as was previously thought; and (3) the "tail first" birth position, seen in the Early Jurassic ichthyosaurs of the Holzmaden deposit (figs. 5.5 and 5.6) and which is like that among sirenians and cetaceans, was a derivative condition, one that arose secondarily and in connection with a pelagic lifestyle.

Locomotion in Ichthyosaurs and Sauropterygians

We can determine the swimming style of numerous species of Mesozoic marine reptiles simply on the basis of their anatomy.

Ichthyosaurs

Ichthyosaurs came in very different body types over the course of the Mesozoic (fig. 5.9).

The most ancient forms (from the Lower Triassic), such as *Grippia* and *Chaohusaurus*, were small, with a long body and a lengthy, narrow caudal fin. They lived in a coastal environment and must have adopted a subcarangiform style of swimming (see "Adaptations to the Aquatic Environment," pp. 118–19).

Forms from the Upper Triassic, such as *Mixosaurus* and *Shonisaurus*, were equipped with an asymmetrical caudal fin (the upper lobe being hardly developed at all) that was compressed laterally. They must have employed a carangiform swimming style.

Among later forms, such as *Ophthalmosaurus* (fig. 5.19) and *Platypterygius* (from the Lower Cretaceous), the caudal fin was homocercal (i.e., the upper lobe was identical to the lower lobe), and the animal's trunk was short and bulky. Propulsion probably was thunniform, accomplished by side-to-side motion of the bilobed caudal fin. The limbs were not used to provide thrust, although they probably were important for stabilization and allowed for rapid changes in direction. These advanced ichthyosaurs lived in the open ocean. Their strong overall resemblance to tuna, certain sharks, and today's dolphins seems to suggest that they were very fast swimmers (the comparison with dolphins is imprecise, because dolphins' caudal fin is oriented horizontally, not vertically). Certain species of tuna, sharks, and dolphins can maintain, over relatively short distances, speeds on the order of 50 kilometers an hour; therefore, ichthyosaurs probably could too.

▼ Fig. 5.9. Diagrams illustrating the similarities in shape between ichthyosaurs and sharks: primitive ichthyosaurs, such as *Chaohusaurus*, had a long body and a long, thin caudal fin; more evolved forms, such as *Stenopterygius*, with a short and stocky trunk and a bilobed caudal fin, were rapid thunniform swimmers.

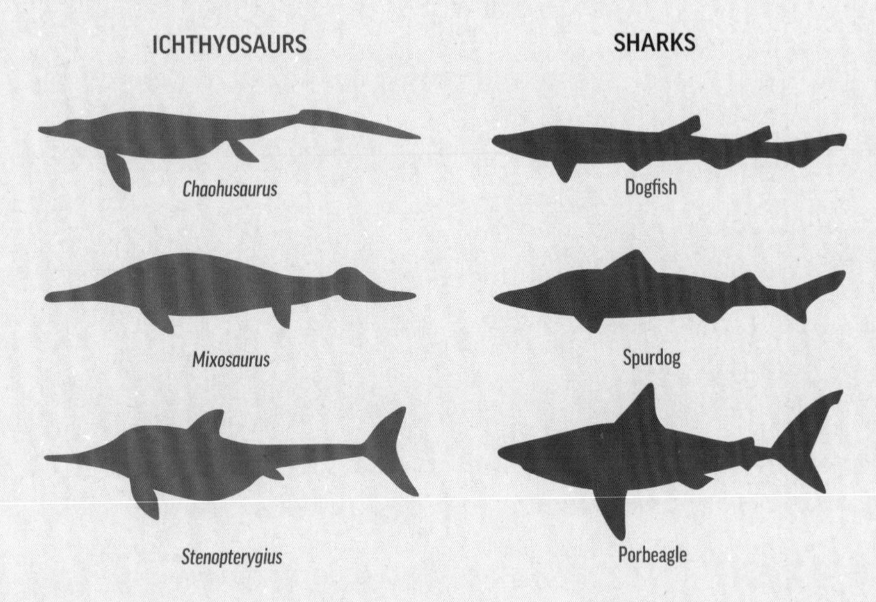

Fragments of fossilized ichthyosaur skin—for example, in the Holzmaden specimens (see p 138)—appear to confirm this hypothesis. The striated structure of these skin samples seems to correspond to two layers of fibers, arranged in different directions. This kind of fibrous organization has been observed in the skin of today's sharks and would seem to endow the skin with elastic properties, allowing it to stretch and flex in a way that limits the emergence of folds in the skin as the animal swims, thereby minimizing turbulence and friction and enhancing acceleration and speed.

Sauropterygians

The marine reptiles of Sauropterygia, because of their extremely varied shape and size, swam in various ways, but their methods were profoundly different from those of ichthyosaurs. Whereas ichthyosaurs' limbs and girdles were used only for stabilization and changes in orientation, sauropterygians' played an important role in propulsion.

The most ancient Triassic sauropterygians, such as the pachypleurosaurs, had a long neck, a long trunk, and a long tail. Their limbs were similar to those of terrestrial vertebrates, except that many forms had longer **autopods** (the parts of the limbs corresponding to hands and feet). These animals must have swum principally by undulating their body laterally (axial locomotion) via movements of their forelimbs and hind limbs (paraxial locomotion) (see "Adaptations to the Aquatic Environment," pp. 118–19). The Triassic sauropterygians were mostly coastal, but their only very slightly modified pentadactyl (five-fingered) limbs could perhaps still have allowed them to move about on dry land. (This last point is still very much the subject of debate.)

▲ Fig. 5.10. Four hypothesized swimming techniques in plesiosaurs.

Among more-evolved sauropterygians, represented by the plesiosaurs, there was increased involvement of the limbs (which had been modified into swimming paddles) and the girdles; it seems that these creatures propelled themselves principally by the use of their swimming paddles (this is known as paraxial swimming). Plesiosaurs' relatively short tail probably was not used for propulsion but was eventually involved in directional control.

Plesiosaur locomotion has been studied in detail by many researchers but is still much debated. Three hypotheses as to the use of their swimming paddles have been advanced: (1) they used their paddles in the manner of oars, the "blades" being used mostly along a horizontal plane from front to rear (fig. 5.10A); (2) they used their swimming paddles as sea lions do, creating thrust via a downward and backward motion (fig. 5.10B); and (3) they performed a figure-eight movement with a cadence comparable to that of underwater flight: the "blades" moved principally along a vertical plane while constantly being turned on their axes so as to generate thrust, in a manner roughly analogous to that of penguins or sea turtles (fig. 5.10C). Some form of consensus does seem to be emerging: plesiosaurs must have engaged in a form of "double underwater flight," in which they moved their four limbs alternately through a series of stages and in which propulsion was provided at the stage where the limbs were lowered (fig. 5.10D). Plesiosaurs were the only vertebrates ever to use both their forelimbs and their hind limbs for propulsion in the water—sea lions and sea turtles use only their forelimbs.

Those plesiosaurs with a long neck and a small head (plesiosauroids) must have moved at relatively modest speeds, since their design was not hydrodynamic. Those with a large head and a short neck (pliosauroids), which were considerably sleeker in shape, must have been more rapid swimmers, and their activity must have been comparable to that of today's large carnivorous marine mammals.

Diving and Underwater Vision

The resemblance between Mesozoic marine reptiles and modern marine mammals extends to their being able to dive to great depths. Some plesiosaur fossils show symptoms of osteonecrosis, a death of bone cells caused by a restriction of blood supply. Dives that were deeper or longer than these plesiosaurs' physiology reasonably allowed for could be the origin of such symptoms. Diving capabilities have also been proposed for at least some species of ichthyosaurs, such as *Ophthalmosaurus*. Judging by the sclerotic ring—a bony ring integrated into the eyeball and composed of overlapping ossicles (tiny bones) (fig. 5.11)—*Ophthalmosaurus*'s eyes, in proportion to its body size, were the largest of the entire animal kingdom (see chapter 2, p. 34). This points to an exceptional visual capacity, which, to these animals that probably could not rely on echolocation as cetaceans can, may have been indispensable during dives to great depths, where light was scarce.

Sclerotic rings are present in several groups of marine reptiles: ichthyosaurs, plesiosaurs, and mosasaurs. They have often been interpreted as adaptations to an aquatic environment, and notably they play a role in both resistance to water pressure and in acting as the focal length of a camera does in vision. But their absence among marine mammals (e.g., cetaceans), and, conversely, their presence among numerous terrestrial reptiles, both living and extinct (i.e., lizards, dinosaurs, pterosaurs, and birds) (fig. 5.11) suggest that they mostly served to support the eyeball, which often reached considerable size in these animals. Among the squamates, for example, studies have shown that the number of ossicles and their arrangement varied among mosasaurs and sometimes resembled and sometimes did not resemble the pattern in some of today's lizards. This great variability suggests that neither the number of ossicles nor their arrangement played a major role in the adaptation of mosasaurs' eyes to the aquatic environment.

The exceptionally large, forward-facing eye sockets of the mosasaur *Phosphorosaurus*, discovered in Japan in 2015, has led to a hypothesis according to which at least some of these animals were endowed with binocular vision, because each eye's field of vision would have overlapped the other's by an estimated 35 degrees (fig. 5.12). Considerably wider than the overlap in field of vision among today's lizards, which

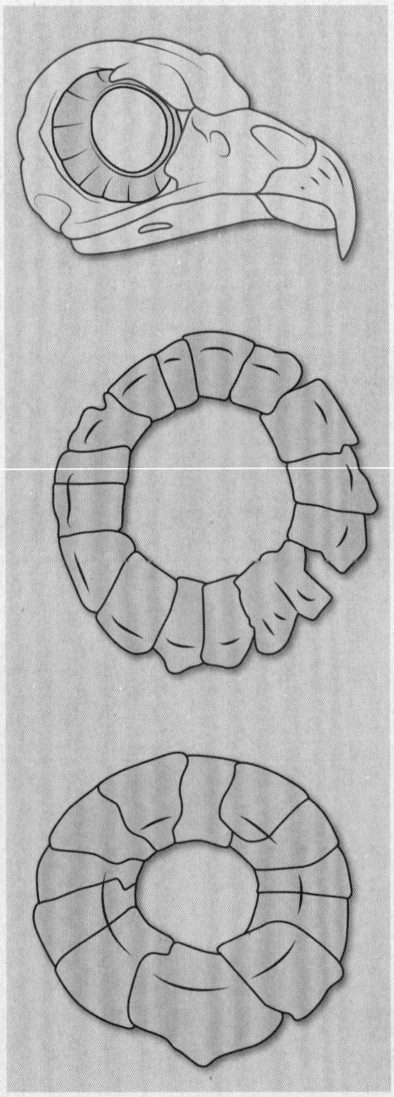

▼ Fig. 5.12. Differences in the lateral arc of overlapping fields of vision among mosasaurs. The size and orientation of the orbits in *Phosphorosaurus* indicate that it could view a significant range of its environment with both eyes at once, suggesting binocular vision.

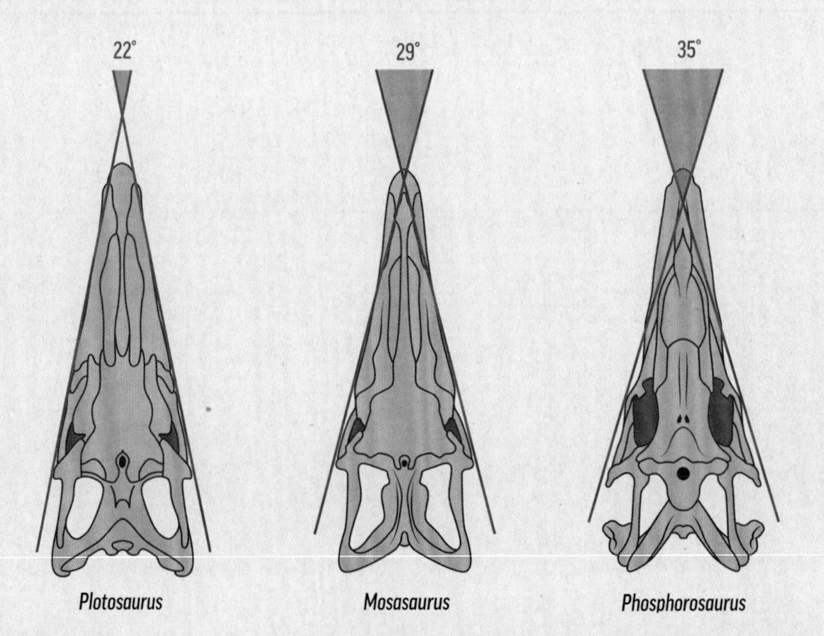

▲ Fig. 5.11. Sclerotic rings in a barred owl, an ichthyosaur, and a mosasaur.

ranges from 10 to 20 degrees, it resembles that of nocturnal snakes. The joint presence in the same deposit of lantern fish and coleoid cephalopods (soft-bodied mollusks, like cuttlefish), two groups that today are generally bioluminescent, indicate that *Phosphorosaurus* may have been adapted to a nocturnal way of life. It probably chased small bioluminescent prey at night thanks to its special vision (fig. 5.14), to avoid direct competition with and the danger of significantly larger predators such as *Mosasaurus*, known from the same deposit.

The study of the endocranium (the cast of the inside of the skull) of marine reptiles, made possible by X-ray tomographic imaging of fossilized skulls, can teach us a lot about their neuroanatomy as well as their sensory capabilities—notably vision. In the case of plesiosaurs (fig. 5.13), the endocranium is seen to be rather long, with long olfactory pathways, relatively small olfactory bulbs, very distinct and developed optical lobes, and a large cerebellum. Plesiosaurs therefore seem to have relied more on vision than on smell and to have swum and hunted in quite particular ways, which likely helped them coexist with other large marine predators.

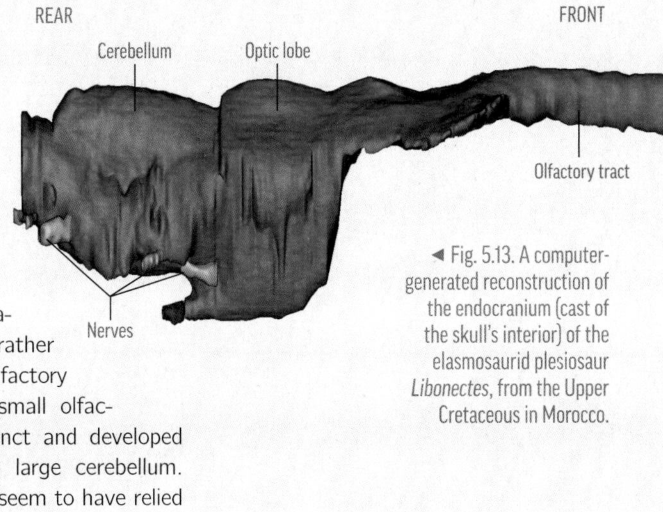

◄ Fig. 5.13. A computer-generated reconstruction of the endocranium (cast of the skull's interior) of the elasmosaurid plesiosaur *Libonectes*, from the Upper Cretaceous in Morocco.

▼ Fig. 5.14. *Phosphorosaurus* chasing some coleoid cephalopods.

III

The Oxford Clay

From the Bottom of the Oceans to the Walls of Houses

The English brickmaking industry is one of the most significant in the world. For its raw material, it makes use of clays deposited in the Middle and Late Jurassic (from the Callovian to the Oxfordian age, 166–157 Ma) from the geological formation known as the Oxford Clay (fig. 5.15), a large band of sediments that crosses the country, from Dorset in the southwest to Yorkshire in the northeast (fig. 5.16). These clays—which, contrary to their name, are exploited mostly in the vicinity of Cambridge, not Oxford—have several properties that brick manufacturers value: they are relatively uniform over large sections of the sediment band and contain a significant proportion of organic matter (about 5%), with a high calorific value, thus allowing for lower firing temperatures, meaning bricks can be produced at lower cost.

▼ 5.15. An open pit quarry in the Cambridge area where the Oxford Clay is mined.

In the mid-nineteenth century, the industrial revolution created a strong demand for bricks, and it was therefore during these years that the quarries were most heavily exploited and a huge number of often complete skeletons of ichthyosaurs, plesiosaurs, and thalattosuchians were discovered. Until the 1920s these quarries were excavated by hand, which aided in the assembling of very important paleontological collections now on display in museums in England and around the world. The subsequent mechanization of extraction made discoveries much more difficult. In addition, the use of other building materials has reduced the demand for bricks, leading to the closure of many brick manufacturers that once exploited the Oxford Clay.

It is interesting to note that this geological formation, composed of sediments deposited in a shallow sea that covered a large part of Great Britain and the north of France, naturally has its equivalent on the other side of the Channel, in Normandy and northern France. Along this French coastal fringe, the sediments occasionally emerge as outcrops in the vicinity of cliffs by the edge of the sea (as they do in Villers-sur-Mer, in the Calvados region). They also, as in England, emerge inland as a result of quarrying. In France, however, quarrying has been much less frequent and the outcrops less extensive; consequently, the fossils unearthed there are both rarer and less complete, although this does not diminish French paleontologists' interest in them! To cite just one example, it was in the layers that correspond to the Oxford Clay in the region of Boulogne-sur-Mer (Pas-de-Calais) that Henri-Émile Sauvage described the typical tooth of the famous pliosaur *Liopleurodon ferox* (see "Top Predator," p. 158). At this French location only this unique tooth, so characteristic, was discovered, while on the other side of the Channel complete *L. ferox* skeletons were exhumed! But the genus and species names *Liopleurodon ferox* will remain associated with the site near Boulogne where the fossil used for the description and naming came from.

If we could take a dip in the Middle Jurassic sea over Great Britain, it would be much more pleasant than a swim in today's Channel; the temperature ranged from 20°C to 29°C, compared to our chilly 7°C–16°C. The accumulation of sediments in this basin was discontinuous: there were periods of heavy sedimentation, in which rapid burial allowed for the preservation of vertebrate skeletons, and long periods of weak sedimentation, represented by the accumulation of layers with shells. The large amount of organic matter in the layers in which vertebrates have been found points to an environment that was significantly oxygen-deprived (anoxic). The conditions that led to the preservation of the numerous fossils in the Oxford Clay must therefore have been like those that prevailed in the Holzmaden region (see p. 138).

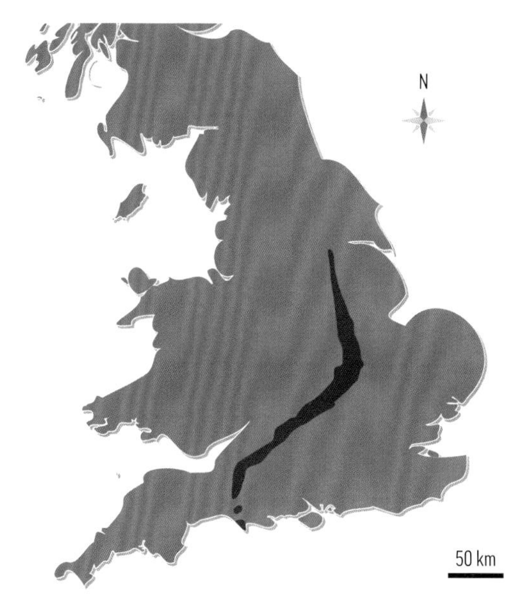

◄ Fig. 5.16. A map showing the extent of the outcroppings belonging to the Oxford Clay in Great Britain.

Ammonites, Belemnites, Giant Fish …

The Oxford Clay deposits provide an exceptional overview of the benthic and pelagic marine fauna from the Middle and Upper Jurassic in Europe. Mollusks are extremely well represented, with around sixty species of bivalves, some gastropods, and close to eighty species of ammonites in an exceptional state of preservation, from the point of view of both their anatomy and their chemical composition. Actually, the composition of their shell—notably their nacre (the lustrous inside lining)—was not transformed by **diagenesis**, and geochemical studies have allowed researchers to determine the paleotemperatures of the oceans and the living environments of the different species. Additionally, the significant number of specimens of certain species of ammonites has allowed for advanced and precise morphometric studies and, for the first time, the highlighting of sexual dimorphism (differences between females and males) in certain species of this group. This sexual dimorphism is seen especially in differences in size and ornamentation.

It is only in very exceptional circumstances, such as those surrounding the creation of these deposits, that fossilization preserves soft tissues or keratin. But in the Oxford Clay, we find the mouth parts of ammonites, their "beak" (aptychus), preserved sometimes in isolated form and, very rarely, in its correct position at the shell's opening. Belemnites, another extinct group of marine cephalopods, did not have an external shell as ammonites do but (as mentioned in our discussion of Holzmaden) had an internal "bone," called a rostrum. These structures, which fossilize well, are often the only parts of belemnites we find. The reason the Oxford Clay can be considered a Lagerstätte is because of its belemnites that were preserved in their entirety, soft tissues included! These exceptional fossils reveal the presence, as in squids, of a mantle and head to which the belemnite's arms, equipped with hooks to capture prey, were attached. Some

They Were Warm-Blooded!

For a long time, Mesozoic marine reptiles' physiology was thought to be similar to that of modern reptiles. For this reason, no questions were ever asked about their metabolism, and by default they were considered **ectothermic** (cold-blooded). Today, most marine reptiles live in shallow environments at latitudes between 40 degrees south and 40 degrees north. The water here remains relatively warm the whole year round, which allows ectotherms to maintain a body temperature and a level of activity sufficient for their movement and feeding habits.

What intrigued researchers is the fact that certain large marine reptiles (plesiosaurs, ichthyosaurs, and mosasaurs) have been found in deposits corresponding to environments at latitudes above 70 degrees during the Mesozoic. As already mentioned (see chapter 1, p. 17), the worldwide climate during the Mesozoic was warmer than it is today. Nevertheless, the temperatures at the poles, while nowhere near as low as those today, could not have exceeded, on average, 10 to 12°C. This leads us to suppose that the thermophysiology of certain Mesozoic marine reptiles must have been different from that of today's marine reptiles, allowing them to withstand fairly low temperatures. Likewise, how can we explain that some of these large marine reptiles with a worldwide distribution were probably great cruisers, capable of crossing entire oceans while swimming constantly, if we do not presuppose a metabolism more elevated than that of a "lambda" reptile?

Histological studies tend to support the notion that these reptiles could regulate their own body temperature. Microscopic examination of cross sections of bone from modern animals allows researchers to analyze the tissue in detail, particularly the orientation of collagen fibers, the type and degree of vascularization, and the density, number, and size of the bone cells (see "The Secrets of Bone," p. 129). In these regards, there are some clear differences between today's **endotherms** (the warm-blooded) and today's ectotherms, the structure of bone tissue being closely tied to the speed of growth, as well as, indirectly, the animal's metabolism: endotherms (mammals and birds) have a more regular and faster rate of bone growth than ectotherms (fish, amphibians, and reptiles) do.

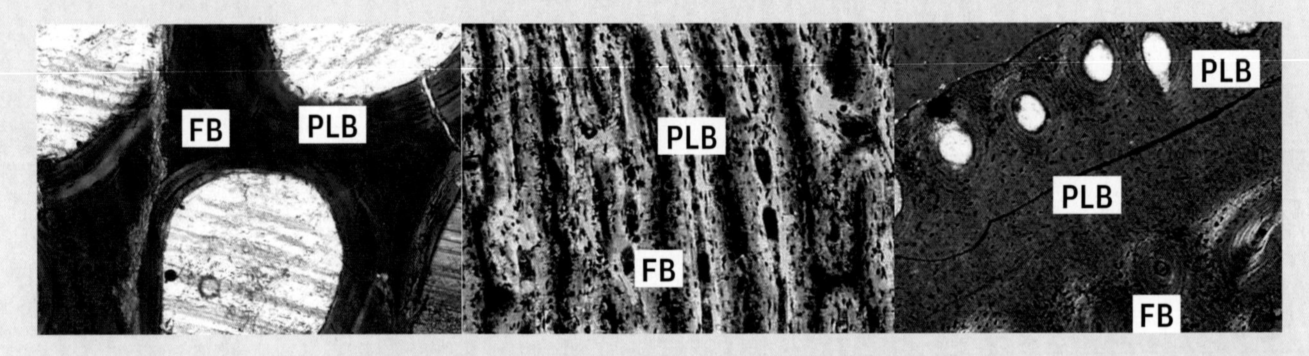

▼ Fig. 5.17. Details from cross sections of bones of marine reptiles that demonstrate the dominance of fibrolamellar tissue with rapid deposition—in other words, fibrous bone (FB)—combined with deposits of pseudo-lamellar bone (PLB) around vascular cavities, in ichthyosaurs (left) and plesiosaurs (center), as contrasted with essentially pseudo-lamellar tissue—in other words, with a slower deposition rate—in mosasaurs (right). These differences in deposition speed essentially translate to differences in growth rate and, therefore, presumably differences at the metabolic level.

close relatives of squids have indeed been found here—once again, preserved to an exceptional degree, with their fine feathers complete (the feather is a "bone" that corresponds to the bone in cuttlefish) and even with traces of ink sacs!

On the vertebrate side of things, the fossils are just as interesting and numerous. The fish are very fragmentary, but their teeth and other parts attest to a large number of species, with widely varying diets and ecological specializations. It is, moreover, from this formation that the greatest number of specimens of *Leedsichthys problematicus* (fig. 5.19), without a doubt the largest bony fish to have ever existed, have come. Known only

Similar studies of ichthyosaur and plesiosaur bones point to rapid growth and therefore, presumably, a physiology of an endothermic type (fig. 5.17). In the case of mosasaurs, the growth rate appears to have been higher than in lizards but significantly lower than in plesiosaurs and ichthyosaurs, suggesting **gigantothermy** (endothermy tied to mass), as leatherback turtles possess. These studies seem to have found some confirmation thanks to an ichthyosaur specimen in which some soft tissues have been preserved, highlighting the presence of a layer of fat beneath the skin. This layer must have worked as insulation, as it does in today's marine mammals (most notably pinnipeds and cetaceans), as well as an energy reservoir and a buoyancy aid.

Some studies using geochemical tools have confirmed the hypothesis that Mesozoic marine reptiles could regulate their own body temperature. The chemical element that was measured in this instance is oxygen, which exists in two distinct atomic forms: ^{16}O and ^{18}O, which are called isotopes because they have the same number of protons but a different number of neutrons. In marine animals the oxygen present in mineralized tissues (i.e., teeth and bones) comes principally from the water they ingested—in other words, seawater. The proportion of ^{16}O to ^{18}O incorporated in the mineralized tissues varies as a function of two parameters: (1) the isotopic composition of seawater (which is more or less rich in ^{16}O or ^{18}O) and (2) the temperature at which the tissues developed.

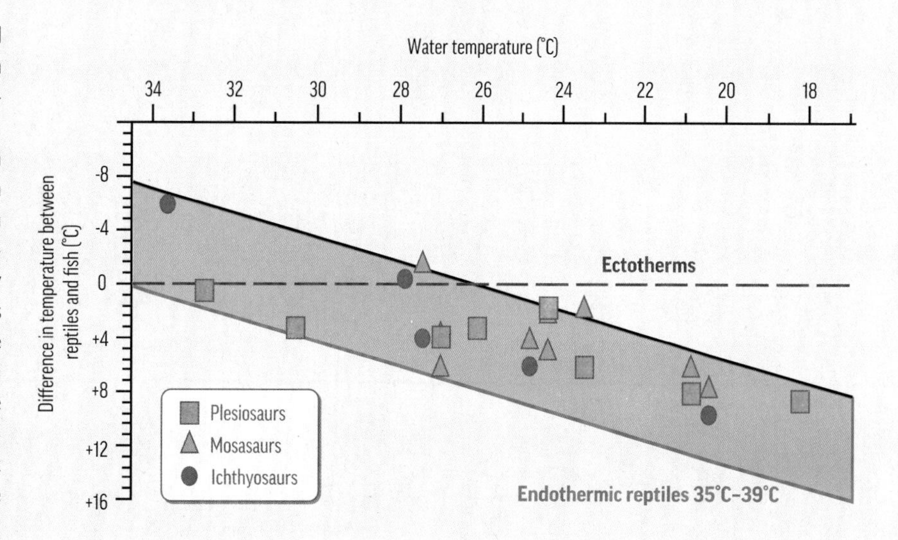

▲ Fig. 5.18. Isotopic variations in oxygen (^{16}O and ^{18}O) highlight a relatively elevated body temperature (between 35 and 39°C) and a constant body temperature in ichthyosaurs, plesiosaurs, and mosasaurs as compared to ectotherms, a fact that points to an elevated metabolism.

For different animals living in the same environment, the isotopic composition of the water they ingest is identical, and therefore only one parameter varies: the animal's body temperature. In fish (which are ectotherms), this temperature is the same as that of the seawater, whereas endotherms' body temperature is relatively constant and generally higher than that of the environment they live in. Analysis of the tissues of endothermic and ectothermic animals that lived in the same waters therefore shows very different oxygen-isotope compositions when the waters they lived in were cold (e.g., at high latitudes).

Analyses of the teeth of plesiosaurs, ichthyosaurs, mosasaurs, and fish that lived in the same marine environment revealed some significant differences in their isotopic composition of oxygen, therefore reflecting differences in body temperature. The marine reptiles' body temperature stayed between 35 and 39°C, even while they swam in waters as cold as 12°C (fig. 5.18). These temperature estimates are close to those of today's cetaceans and suggest that plesiosaurs and ichthyosaurs had an elevated metabolism suited to rapid swimming over long distances, including in cold waters.

from some very fragmentary remains (hence the species name), its length must have exceeded 15 meters—some have estimated it at 30 meters. Luckily for its contemporaries, *Leedsichthys* was a peaceful, plankton-eating filter feeder, like today's largest fish, the whale shark, which can reach a length of 18 meters.

The Plesiosaurs Take Over

So, fish could reach impressive sizes, but marine reptiles weren't exactly in the background either. All the cephalopods and the fish just discussed served as food for a whole array of marine reptiles, from the smallest to the largest.

▲ Fig. 5.19. (Preceding pages) A Middle to Late Jurassic seascape, Cambridgeshire (England). The carcass of a giant actinopterygian, *Leedsichthys*, serves as a meal for a pliosaur, *Simolestes* (on the right), and a crocodylomorph, *Tyrannoneustes* (on the left). In the center is a hybodont shark. In the background are two ichthyosaurs, both of them *Ophthalmosaurus* (on the left), and a plesiosaur, *Muraenosaurus* (on the right).

The Oxford Clay has delivered some important plesiosaur fossils, especially in its lower part. They are most often found in isolation. The group to which each belongs is easily identified, but determining genus and species is rather more difficult. Teeth are one feature that can help. Plesiosauroid teeth were long and narrow, revealing that their diet was more selectively piscivorous than that of pliosauroids, which had more massive and robust teeth and are known as much more opportunistic predators. Within these two groups, tooth surface sometimes was different between one species and another: teeth could be smooth or rough, deeply grooved or shallowly grooved, marked with facets (areas of wear) or not, and so on. The teeth of ichthyosaurs can be easily distinguished from those of plesiosaurs based on more general features, such as an enamel root, markedly pleated dentine (middle layer), and a short crown. As far as the teeth of thalattosuchian crocodylomorphs are concerned, while those of the teleosauroids (which were long and thin) are easily distinguished from those of the metriorhynchoids (which were shorter),

Top Predator: *Liopleurodon*

Name: "teeth with a smooth surface"
Classification: Sauropterygia, Eosauropterygia, Pliosauroidea, Pliosauridae
Lived: Middle Jurassic (Callovian, 165–161 Ma)
Known range: France, England, Germany, Switzerland, perhaps Poland, and probably Russia
Skull length: up to 1.5 meters
Overall length: about 10 meters
Diet: fish, cephalopods, marine reptiles

Liopleurodon was described based on a single tooth, found in the environs of Boulogne-sur-Mer (Pas-de-Calais, Hauts-de-France) in 1873 by French paleontologist Henri-Émile Sauvage. Later, much more complete specimens were unearthed in France, England, Switzerland, and Germany. *Liopleurodon* is monospecific—*L. ferox* is the only species—and yet this is without a doubt the most widely known pliosaur. This ferocious predator, with a large and long muzzle and powerful jaws, presumably tackled sizable prey, such as sharks and other marine reptiles. Its teeth were up to 20 centimeters long, nearly the size of *Tyrannosaurus rex*'s. *Liopleurodon* was therefore the largest reptile and the second-largest creature in the oceans of the Middle Jurassic. Only the actinopterygian fish *Leedsichthys*, which lived in the same waters, exceeded it in size, with a length of more than 15 meters (see fig. 1.9, p. 19).

► Fig. 5.20. *Liopleurodon*, a pliosaurid from the Middle Jurassic in Europe, in pursuit of a small plesiosaur.

in the case of the latter an in-depth study has led to the determination of a number of characteristics, such as the features of the enamel and the shape of the carina (edges), which in combination allow us to differentiate between at least three of the species represented in the Oxford Clay.

Many remains are particularly well preserved, sometimes with a complete and articulated skeleton. The most important discoveries date from the construction of a railroad in the county of Wiltshire in the 1840s. Reports about the first fossils, circa 1843, attracted a number of collectors. The most famous of them, Alfred and Charles Leeds, assembled the most important collections. The first marine reptiles found in the Oxford Clay were the subject of publications in the 1870s and 1880s. Nonetheless, it was the monographic works published by Charles Andrews in 1910 and 1913 that revealed the richness and great variety of marine reptiles to be found here. After winning a position in the Geological Department of the British Museum (Natural History) in London in 1892, Andrews first studied fossil birds, but

following the bequest of the Alfred Leeds collection, he became interested in marine reptiles, producing two monographs in which he described ichthyosaurs, plesiosaurs, and crocodylomorphs. These monographs are still an important reference today for paleontologists!

Although the marine reptiles found in the Oxford Clay are of the same groups as marine reptiles from the Lower Jurassic, the relative proportions of those groups are different, as are the species involved. Plesiosaurs in the wide sense of the term represent more than half the marine reptiles found in the Oxford Clay's Callovian deposits, whereas they are scarcely represented at all in Holzmaden's Toarcian deposits. Their size, which in the Early Jurassic had been modest, reached impressive proportions, with some pliosaurids (e.g., *Liopleurodon*; see "Top Predator," p. 158)—the dominant forms—reaching more than 10 meters long. Other, smaller forms exhibit a variety of ecological adaptations. Take the pliosaur *Peloneustes*, for example, with its long and slender skull equipped with narrow teeth (probably a piscivore), or *Simolestes* (fig. 5.19), with a skull that was short and thickset, armed with robust rounded teeth (probably a durophage). Then there are long-necked plesiosaurs, such as *Cryptoclidus* and *Muraenosaurus*, which likely ate fish as well. These really were the large predators of the Middle and Late Jurassic, and their different morphologies attest to a partitioning of the ecological niches.

On the other hand, the large ichthyosaurs had disappeared, replaced by the slender *Ophthalmosaurus*, with a more compact body but nonetheless up to 5 meters long and probably faster than ichthyosaurs of before. Overall, ichthyosaurs went from being the dominant marine reptiles between 183 and 174 million years ago in Holzmaden to representing less than one-quarter of the marine reptiles between 166 and 157 million years ago in the Oxford Clay.

The other growing presence of the Middle and Late Jurassic was that of the thalattosuchian crocodylomorphs. In addition to the fact that they represent almost a quarter of the reptilian faunas from the Oxford Clay, they were represented no longer only by coastal forms, but also by the metriorhynchoids—animals fully adapted to marine life, equipped with swimming paddles and a hypocercal tail. The metriorhynchoids from the Oxford Clay, such as *Suchodus* and *Tyrannoneustes*, heralded the forms at the end of the Jurassic, with an increase in size, a shorter muzzle, stronger and stronger teeth, and ever more explicit ziphodontics (see "Show Me Your Teeth and I Will Tell You Who You Are!," pp. 180–81). They mark the passage from piscivorous forms with a slender snout and narrow teeth, such as *Metriorhyncus*, to those such as *Torvoneustes* and *Geosaurus* but, above all, *Dakosaurus* (see opposite), which show a tendency toward hypercarnivory—a diet mostly of large vertebrates, such as other marine reptiles, and much less of fish.

In other words, some profound changes occurred between the faunas of the Early Jurassic, like those from Holzmaden, and those of the Middle and Late Jurassic, like those from the Oxford Clay. The groups of marine reptiles were certainly the same, but many of the latter forms were significantly specialized in favor of the open ocean, and dominance had passed from ichthyosaurs to plesiosaurs. And yet even more important changes awaited marine reptiles during the transition from the Jurassic to the Cretaceous.

Top Predator: *Dakosaurus*

Name: "shredder lizard"
Nickname: "Godzilla"
Classification: Crocodylomorpha, Crocodyliformes, Thalattosuchia, Metriorhynchoidea
Lived: Late Jurassic to Early Cretaceous (Oxfordian–Berriasian, 161–140 Ma)
Known range: Europe (Germany, England, France, Switzerland, and maybe Spain), Argentina, and maybe Mexico
Skull length: 80 centimeters
Overall length: 4–5 meters
Diet: fish, cephalopods, marine reptiles

Just like the other metriorhynchoids, the most aquatic of the crocodyliforms, *Dakosaurus* was equipped with both swimming paddles and a fishlike tail, very similar to those of ichthyosaurs. Its teeth resembled finely serrated knife-blades (i.e., they were ziphodontic), unlike most other marine crocodylomorphs' teeth, which were long and pointed with a round cross section and smooth edges in. The latter were generally piscivorous, even though certain species with massive teeth were more opportunistic and didn't hesitate to tackle large vertebrates. *Dakosaurus*'s ziphodontics, as well as its short muzzle, suggest it preferred large prey, such as other marine reptiles. The very high skull of the species *D. andiniensis*—because this shape increases the skull's ability to withstand significant forces—attests to a tendency toward mega-predation.

▲ Fig. 5.21. *Dakosaurus*, a thalattosuchian crocodylomorph from the Upper Jurassic and Lower Cretaceous in Europe and South America.

CHAPTER 6

THE CRETACEOUS PERIOD

First 3D reconstruction of a mosasaur, joint work by the naturalist Richard Owen and the artist Benjamin Waterhouse Hawkins (1854). The Crystal Palace Park, Sydenham, England.

Champagne

The Cretaceous (145–66 Ma), the last period of the Mesozoic era, was defined by the Frenchman Jean-Baptiste d'Omalius on the basis of rocks from the Paris Basin. Its name comes from the Latin *creta*, "chalk," referring to the vast deposits of this substance that characterize this period with outcroppings all over northern Europe. Chalk makes up the subsoil of the entire Paris Basin, where it has often been exploited in underground quarries—as, for instance, in Meudon, where it used to be extracted to make the whiting known as "Blanc de Meudon" (or Paris white, or Spanish white). It constitutes the ideal substrate for a wine region that is celebrated the world over: the Champagne. It also adds its signature to the magnificent cliffs, entirely dressed in white, along the coasts of the Channel and the North Sea (in France, England, Germany, and Denmark) that have provided so much inspiration to painters, from Claude Monet to Caspar David Friedrich. This chalk, formed by the accumulation and deposition of billions of coccoliths—plates of calcium that protected coccolithophores (phytoplankton)—at the bottom of an ancient sea, holds a treasure of fossilized marine reptiles. The Cretaceous ended with a biological crisis, which, although not the most devastating ever experienced, is certainly the most widely known, because it corresponds to the extinction of the non-avian dinosaurs.

▼ Fig. 6.1. An open-air opal mine in Coober Pedy, in the state of South Australia.

In the Cretaceous world, the sea reached its highest level in the history of Earth. The climate was warm worldwide, with temperatures about 4°C higher than they are today (the average temperature was 18°C), and with very high levels of carbon dioxide (CO_2) (six times those known prior to the industrial revolution). Whereas the faunas of the Late Jurassic had been dominated by the plesiosaurs, the thalattosuchian crocodiles, and—to a lesser extent—the ichthyosaurs, starting in the Early Cretaceous this composition slowly changed, until the Late Cretaceous population of marine reptiles was significantly different.

First, during the Early Cretaceous, the number of thalattosuchian species (see chapter 2, p. 63) plummeted, leading to their extinction at the beginning of the Aptian age, roughly 125 million years ago. Then, during the Albian age, the ichthyosaurs—which, during the Early Cretaceous, contrary to widespread prior belief, were still represented by a range of forms (see chapter 2, p. 31)—began to become rarer; during the biological crisis at the end of the Cenomanian (90 Ma) they completely disappeared, although we do not know precisely why. It was during the Cenomanian (ca. 100 Ma) that the squamates took to the seas (see chapter 2, p. 96); their most emblematic group, the mosasaurs, became one of the major representatives of the faunas of marine reptiles throughout the Late Cretaceous. Last, the plesiosaurs were affected by the great faunistic upheavals during this period too: the pliosauroids declined and went extinct before the end of the Cretaceous, while the two remaining groups of plesiosauroids—the elasmosaurids, with their gigantic necks, and the polycotylids, in the shape of pliosaurs—diversified (see chapter 2, p. 62). It was also during this transitional period, from 120 to 100 million years ago, that the first sea turtles—chelonioids and bothremydids—appeared (see chapter 2, p. 81). Mosasaurs, plesiosaurs, and turtles therefore dominated marine ecosystems for the entirety of the Late Cretaceous. The first two groups did not survive the Cretaceous/Paleogene crisis, but that same crisis was a boon for certain crocodyliforms. We shall return to this extinction at the end of this chapter.

The Eromanga Basin

Geographic and Stratigraphic Context

The 1-million-square-kilometer Eromanga Basin (fig. 6.1) in east central Australia comprises portions of Queensland, New South Wales, the Northern Territory, and South Australia (fig. 6.2). In this region, three deposition sequences follow one another: a sequence of continental (fluviatile) origin during the Jurassic is overlaid with deposits of marine origin from the Lower to Upper Jurassic, followed by another sequence of deposits with continental origin. More than a dozen sites in the Eromanga Basin have yielded fossils of marine reptiles. Among them is White Cliffs, in the state of New South Wales. During the Early Cretaceous, this region corresponded to a peninsular area in the east of Gondwana, near Antarctica, at roughly 70 degrees south latitude. White Cliffs is part of the vast deposits of carbonate-rich schists and clays found in outcroppings in the Eromanga Basin today. These deposition sequences attest to a shallow coastal marine environment, possibly with anoxic conditions on the bottom. In White Cliffs the Wallumbilla Formation is where fossil marine reptiles around 120 million years old (from the Upper Aptian) have been found.

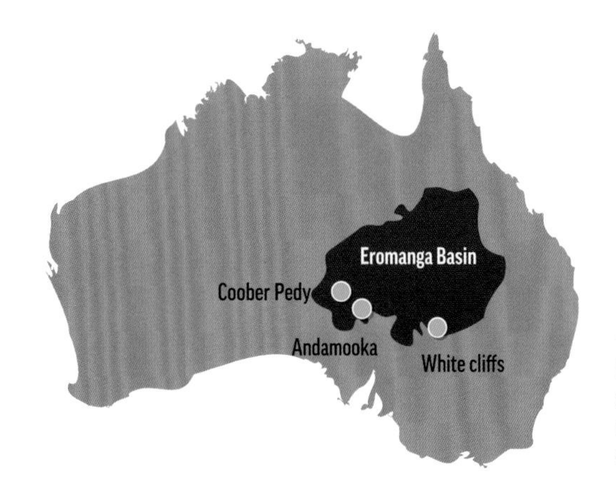

◄ Fig. 6.2. A map of the Eromanga Basin and the main deposits of both opals and marine reptile fossils.

▲ Fig. 6.3. (Preceding pages) An Early Cretaceous seascape, Eromanga Basin (southern Australia). In the center, a *Kronosaurus*, the giant pliosaur, chases an *Umoonasaurus*, a little plesiosaur with astonishing skull ridges. On the bottom left, another plesiosaur, *Opallionectes*, dives toward a school of actinopterygian fish, while on the upper right an ichthyosaur, *Platypterygius*, is about to surface.

Historical Notes

White Cliffs was Australia's first commercial opal field, with intensive opal mining beginning in 1889. The first discovery of marine reptiles in this opal-bearing Cretaceous deposit was reported by English geologist Henry Woodward in 1895. Woodward noted fragmentary remains and a tooth of considerable size, attributed originally to the pliosaur *Polyptychodon* but today ascribed to *Kronosaurus* (fig. 6.3). Other bones, including those of various plesiosauroids, as well as the fragment of an ichthyosaur vertebra, were soon described at the end of the nineteenth and the beginning of the twentieth century.

Today, large-scale opal-mining operations in White Cliffs have mostly ceased, to the extent that there are practically no longer any "official" fossil discoveries. Fossils are regularly found by private individuals, owners of certain mines, but these specimens usually end up in private collections and thus are unavailable for research purposes.

Paleoenvironments

The White Cliffs sediments were deposited during the Early Cretaceous, in a shallow epicontinental sea, during a period of high sea levels. The paleoclimatic indicators for the Lower Aptian/Upper Albian marine deposits are such that the nature of the sediments and the sedimentary objects found there (see below) agree with the isotopic estimates of the sea's paleotemperatures in the southwest of the Eromanga Basin. They point to very cold conditions, with estimated average lows of between 0°C and 10°C, strongly seasonal, with possible freezes in the winter. In the sediments of the Bulldog Shale formation (from the Upper Aptian) are dropstones (clasts the size of a cannonball) as well as glendonites (a type of carbonated concretion). These dropstones were not transported by the usual marine currents but instead released vertically into the water column. Their presence in these fine sediments may be a result of their being displaced by the movement of glaciers on land, integrated into the ice, then transported into the sea across long distances inside icebergs. Once these melted, they would have been released and the dropstones would have sunk to the ocean floor. As far as glendonites are concerned, they are known to form in waters close to freezing. These deposits are associated with a distinctive fauna of invertebrates, but also of vertebrates, especially marine reptiles.

Taphonomy and Paleontology

Fossils of marine reptiles found at White Cliffs generally consist of partially articulated skeletons or isolated elements, suggesting that their remains drifted in the sea for a while before settling to the bottom. The mode of preservation seems indicative of a shallow environment and of a possibly anoxic seafloor.

Because it occurs in direct association with the indicators of a cold climate we mentioned above, the presence of marine reptiles in the assemblage of fossils at White Cliffs (and certain other areas in southern and southeastern Australia) is peculiar. The climate of the high southern latitudes at which these creatures lived is a stark contrast to the generally warm climates favored by reptiles today, and it shows that certain Mesozoic reptiles, such as the ichthyosaurs and the plesiosaurs, were able to survive extremely low temperatures. These creatures must have had some specific physiological or behavioral adaptations that allowed them to conquer cold-water environments (see "They Were Warm-Blooded!," pp. 154–55). Notably, Cretaceous turtles and crocodiles that do not exhibit any of the same physiological adaptations are not found in these high-latitude deposits.

Among the marine reptiles that have been inventoried at White Cliffs, the plesiosaurs were by far the most diversified (there were

◄ Fig. 6.4. Detail of the plesiosaur *Umoonasaurus* from the Lower Cretaceous in Australia, showing the peculiar bony projections atop its snout and its brows.

three taxonomic families) and the most abundant. This group also dominated the faunas of Late Cretaceous marine reptiles that have been found worldwide at high paleolatitudes. The composition of these faunas strongly contrasts with that of the faunas of the same epoch at low paleolatitudes, in which plesiosaurs were generally much less abundant and diversified when compared to, for instance, the mosasaurs. The reasons for this difference are not entirely understood, but they could be tied both to these plesiosaurs' greater tolerance for cold environments and to their occupation of different ecological niches in order to minimize competition. The absence of other marine reptiles, such as turtles, would also attest to the barrier to their dispersal represented by these cold waters.

The fauna of marine reptiles at White Cliffs underscores the global distribution of the ichthyosaurs and plesiosaurs. For example, the ichthyosaur *Platypterygius* (fig. 6.3), known from White Cliffs, is found the world over, and the large pliosaur *Kronosaurus* is found both in Australia and in South America. Yet the faunistic assemblage at White Cliffs includes some potentially isolated forms too—specifically, the oldest known polycotylids, suggesting that this taxonomic family originated in these high-latitude epicontinental seas in Gondwana's east during the Early Cretaceous. If that is true, the polycotylids must have spread and very significantly diversified in the northern hemisphere over the course of the mid-Cretaceous (ca. 100 Ma), as their remains are known from the Upper Albian in North America and, starting in the Cenomanian, from Asia and Europe.

Fossils in Opal

The processes of mineralization of organic remains during their fossilization depend on biochemical and physico-chemical changes that affect the sediment surrounding them. The various fluids that circulate in the sediments around the fossils notably play a fundamental role. These fluids can either partially or entirely dissolve the mineralized parts of the remains (e.g., bones, teeth, or shell), or replace them with new minerals while maintaining the initial mineral structure (in a phenomenon known as crystallo-chemical substitution), or even make the initial mineral structure disappear while preserving the original chemical composition. Replacement of mineralized tissues with opal (an amorphous hydrated silicate, $SiO_2 \cdot nH_2O$, a mineral that is used as a gem) is, rather straightforwardly, called opalization.

Australia produces more than 90% of the world's opal gemstones—used mostly in jewelry—and they are exploited either in the form of veins or in the form of crystallized nodules in layers of clay, schists, and sandstone. The main opal mines are in the Eromanga Basin. As this mineral was forming in the area, some ichthyosaur and plesiosaur fossils underwent opalization. Several either partially or completely opalized skeletons have been discovered in various deposits of the Eromanga Basin, most notably at Andamooka, Coober Pedy, and White Cliffs (figs. 6.1 and 6.2).

Even so, not all the fossils in these deposits are opalized. The processes leading to their opalization are still poorly understood, and some of the main questions concern (a) the age of the replacement of the biomineral by the opal, (b) whether the silica within the opal was of sedimentary or volcanic origin, and (c) the composition of the opalizing fluids. It would seem, however, that the opal was first deposited in gel form in the bones' micro-cavities and, from there, extended to either part of or the entirety of the bone. Changes in the viscosity of the gel determined whether the opal would be colored. Their transformation into opal notwithstanding, these Australian fossils partially preserve the microstructural characteristics of bones, even down to the histological level!

The most complete opalized marine reptile fossil is of an *Umoonasaurus* (fig. 6.3), a plesiosaur about 2.5 meters long with strange bony ridges atop its snout and above its eyes (fig. 6.4). This specimen, doubly precious because of its opalization and its peculiar features, was discovered by a miner in Coober

Top Predator: *Kronosaurus*

Name: "lizard of Kronos"
Classification: Sauropterygia, Eosauropterygia, Pliosauroidea, Pliosauridae
Lived: Early Cretaceous (Aptian–Albian, 125–100 Ma)
Known range: Australia and Colombia
Skull length: about 2 meters
Overall length: roughly 10 meters
Diet: fish, cephalopods, marine reptiles

This animal, was first described in 1924, based entirely on a mandible in very fragmentary condition but of astonishing proportions, which led the author, Heber Albert Longman, to estimate the overall size of the teeth at 25 centimeters. A magnificent specimen, found in 1931 by a Harvard group on a mission to Australia, was entirely prepared and put on display in 1959 at the Harvard Museum of Comparative Zoology. The animal mounted in this form, a third of which was reconstructed in plaster, is 12.8 meters long, and its skull alone is 2 meters in length! However, the total number of vertebrae of this specimen of *Kronosaurus queenslandicus* is uncertain, and its overall length was probably overestimated. This mega-predator probably never exceeded 10 meters in overall length, which is still very large!

Kronosaurus was equipped with a long, slender rostrum and relatively short neck, composed of twelve vertebrae. It is the only genus of pliosaur from the Cretaceous in Australia and was one of the largest pliosaurs in the world. Another species, *K. boyacensis*, has been described from the Aptian stage in Colombia, a fact that attests to this animal's great capacity for dispersion.

► Fig. 6.6. *Kronosaurus*, a pliosaur from the Lower Cretaceous in Australia and Colombia.

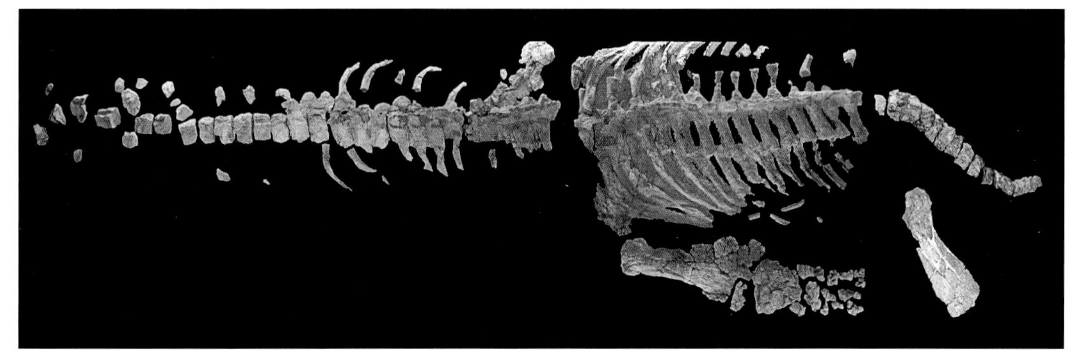

◄ Fig. 6.5. Skeleton of the plesiosaur *Opallionectes*, entirely opalized, South Australian Museum (Adelaide, Australia).

Pedy during the 1980s. Today, it is part of the collection of the Australian Museum in Sydney.

Another plesiosaur, about twice as long as *Umoonasaurus*, is known from a single specimen discovered in an opal mine near Andamooka (fig. 6.2), in the Bulldog Shale formation. This fossil dates from the Aptian or the early Albian age. Consisting of a good deal of skeleton and some teeth, it is entirely opalized (fig. 6.5). The genus was named *Opallionectes*—from *opallios*, "opal," and *nektes*, "swimmer."

III

Morocco's Phosphates

The Ocher Gold from the Arab Provinces

The mining and fossil-bearing deposits tied to Morocco's phosphates are superlative. Their exceptionality is probably due to the triple combination of their privileged paleogeographic position, at the junction of the Tethys and the Atlantic; the specific circumstances of their formation; and the upwellings in the sea at this location.

Morocco's phosphates are part of an immense band of sediments between 20 degrees north and 20 degrees south latitude that stretches from the Middle East, across North and West Africa, all the way to eastern Brazil. So, during the Late Cretaceous (and part of the Paleogene), while chalk sediments were being deposited at essentially the bottom of the seas that covered Europe (see p. 164), different physico-chemical conditions favored the deposition of phosphate sediments in the seas at the edges of the African Craton.

▼ Fig. 6.7. A phosphate quarry in the Khouribga region, Morocco.

These sediments were deposited in a shallow gulf that covered a large part of Morocco, called a "phosphate sea" (fig. 6.8). This process started around 70 million years ago, in the Maastrichtian age (the end of the Cretaceous), and continued in a relatively uninterrupted fashion until around 45 million years ago, in the Lutetian age (the beginning of the mid-Eocene)—in other words, for 25 million years. Of all the phosphates from northwest Africa and the Middle East, these are the ones with the amplest stratigraphic extension. This shallow ocean gulf developed while temperatures were warm worldwide and was an area of plentiful upwellings, a fact that partially explains the wealth of vertebrate fossils (see further on). These sediments are known from several regions of Morocco, notably the mining areas of the Ouled Abdoun and Ganntour basins (fig. 6.8).

From Iraq through Syria, Jordan, Egypt, Tunisia, Togo, and all the way to the state of Pernambuco in Brazil, the sediments are exploited for economic reasons. The phosphates they contain are, in effect, a precious manna used to produce fertilizers and their derivatives (phosphates are an organic compound of phosphoric acid). This mining activity is particularly intensive in Morocco and puts the country in first place as a phosphate exporter. In fact, Morocco's output of almost 25 million tons a year represents about 30% of global phosphate production. Its phosphate reserves alone are estimated to constitute around three-quarters of worldwide reserves.

Since its creation in 1920, Morocco's Office Chérifien des Phosphates (which in 2008 became the OCP Group) has managed the whole process, from the extraction of phosphate sediments to their transformation into derivatives (fertilizers, phosphoric acid, etc.). The machinery used for extraction in

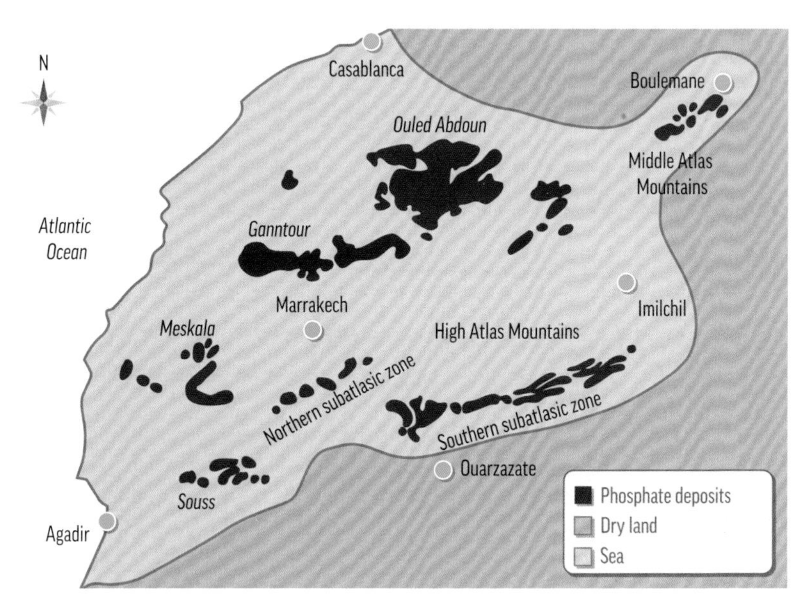

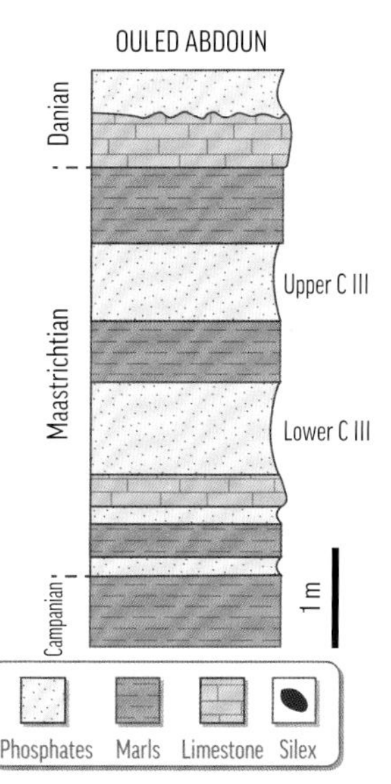

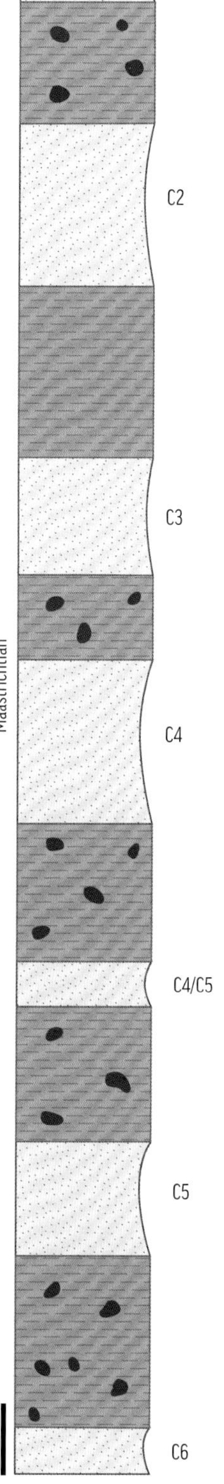

▲ Fig. 6.8. A map showing the principal phosphate deposits in Morocco (dark brown), over a paleogeographic map showing how far the "phosphate sea" extended at the end of the Cretaceous. The principal mining areas are the Ouled Abdoun and Ganntour basins, where the Cretaceous sedimentary series have vastly different breadths: 4 meters at Ouled Abdoun, compared to 22 meters at Ganntour! The different layers that are being exploited are indicated by a C in the stratigraphic logs.

the active mining areas reflects the economic might of the OCP: there are dredges as large as buildings; trucks with wheels that are 4 meters in diameter; bulldozers so huge that their operators must get in and out by ladder; strange, flattened drills that look like crawling insects. ... All this mechanical activity tirelessly—24/7, 365 days a year—crisscrosses a lunar-looking landscape made up of innumerable tracks intersecting endless trenches bordered by huge slag heaps (fig. 6.7). Together with tourism and remittances from Moroccans living abroad, the OCP is a top source of currency flows into Morocco.

A Franco-Moroccan History

Morocco's phosphate deposits were first discovered in 1905 by French geologist Abel Brives, while he was prospecting between Essaouira and Marrakech. But the interest aroused by the discovery was, for the most part, limited to geologists, who argued over the deposits' stratigraphic age. It was the 1917 discovery of gigantic phosphate deposits near the town of El-Borouj, to the south of the Ouled Abdoun Basin, that sparked the exploitation of phosphates in Morocco. French authorities (from 1912 to 1956, most of Morocco was a French protectorate) quickly seized control of the economic and industrial interests associated with the immense deposits. (And the exploitation of these Moroccan phosphates partially triggered the exploitation of the Quercy phosphorites in France.)

Starting in 1920, phosphate was mined underground in the center of the Ouled Abdoun Basin, but this practice rapidly gave way to open-pit mining. The exploitation then spread to other basins, such as Ganntour, where exploitation started in the 1930s but really took off in the 1980s. Geologists, however, soon ran into dating and correlation problems. The phosphate sediments all looked very similar, and any differences that were apparent could be misleading.

A layer in one basin could be the same color as a layer of a different age in another basin just a short distance away, and layers of the same age could be different colors. Moreover, for reasons tied to physico-chemical reactions during the creation of these phosphates, the usual fossil markers (ammonites, foraminifera, etc.—creatures with shells and other hard parts made of calcite, $CaCO_3$) were preserved very poorly in those instances when they were preserved at all, making it impossible to date the different layers in a traditional reliable fashion. Under these conditions, how could scientists carry out a rigorous study of the deposits being mined? It just so happened that the teeth and bones of fossil vertebrates, made of hydroxyapatite (phosphate, in other words), were very well preserved and incredibly abundant in these deposits. This was especially the case with the teeth of selachians (sharks and rays)—which paleontologists had used as index fossils in some locations since the nineteenth century, the same way they relied on ammonites and microfossils in other locations, to help them identify **biozones** and thus date the layers.

It was in this context that French paleontologist Camille Arambourg was given a mandate, starting in the early 1930s, to study the successions of selachian fauna present in the different basins and phosphate layers of Morocco, Algeria, and Tunisia, in order to establish a precise and reliable stratigraphy that would assist in the exploitation of the mines. Between 1935 and 1950, Arambourg collected and studied more than ten thousand specimens! The book he published in 1952 is still the best reference in existence for any paleontologist working on the phosphates of these regions.

After Amamboug's pioneering work, this paleontological heritage remained scientifically underutilized (except for the selachians), until the 1996 revelation of fossils of terrestrial mammals in these marine phosphate deposits led to the discovery of both incredibly rich and exceptionally interesting vertebrate faunas that made this area an almost unrivaled "hot point" of biodiversity during the Cretaceous/Paleogene crisis.

▼ 6.9. An *Enchodus lybicus* tooth (about 5 centimeters long) and a reconstruction of this large teleost fish from the Upper Cretaceous in Morocco. Tooth, OCP Group (Khouribga, Morocco).

Ali Baba's Cave

The sedimentary "book" of Morocco's phosphates preserves successive pages of the evolution of life on Earth within an almost continuous narrative—albeit intersected by the famous Cretaceous/Paleogene boundary—as a precious record. It constitutes a unique testimony to the teeming life that filled the epicontinental seas on the borders of the West African Craton and its continental countryside for 25 million years.

The fossils in Morocco's phosphates represent every group of vertebrates, from cartilaginous fish all the way through bony fish and reptiles (birds included) to mammals. Only the amphibians are absent, but this is not surprising, since these animals generally do not live in a marine environment. Almost the entirety (96%) of the species inventoried—more than 350!—were marine, but occasionally the rare continental creature, the remains of which likely floated out to sea, attests to the life of the continental countryside both at the end of the Cretaceous and during the Paleogene. The latter sorts of fossils include non-avian dinosaurs, pterosaurs, and mammals.

The chondrichthyan fish (or selachians) are the most abundant and diversified group in these phosphate series, with close to 250 species of sharks and rays of all sizes (from several dozen centimeters to 6 meters in length) and of all ecological types (filter feeders, "crushers," scavengers, and mega-predators) (fig. 6.10). Perhaps the best known among them is the lamniform (mackerel) shark *Scapanorhyncus*, almost 5 meters long, from the Maastrichtian. Then there is the ray *Rhombodus* (an excellent dating indicator for the Maastrichtian), with its astonishing dental pad composed of rhomboid-shaped teeth (hence the name *Rhombodus*) for crushing prey. These selachians, whose cartilaginous skeletons are rarely preserved, are represented essentially by their enameled teeth (fig. 6.10), which number in the billions in these sediments. As previously pointed out, these teeth allow scientists to date the terrains just as reliably as ammonites or microfossils could.

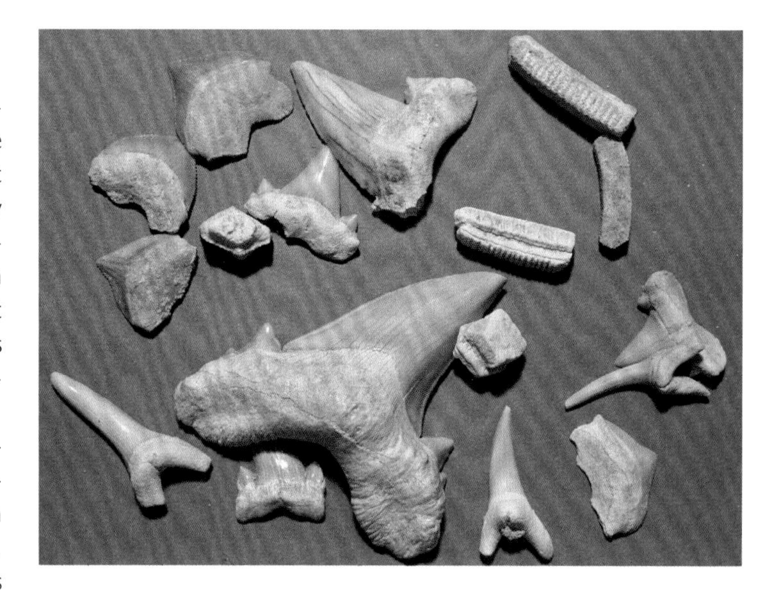

▲ Fig. 6.10. This sampling of selachian teeth from the phosphates of Morocco, including the teeth of rays (flat) and sharks (pointy), points to a great diversity, both systematic and ecological. The largest tooth, about 4 centimeters long, is that of *Otodus*, an extinct mackerel shark.

The actinopterygian fish, for their part, are represented mostly by teleosts and pycnodonts, fish such as *Phacodus*, with teeth meant for crushing. In some ways, pycnodonts resembled today's parrotfish, which browse on coral.

The teleosts are very abundant in all the layers, and the small ones, in the manner of today's schools of anchovies or sardines, must have constituted the sort of immense biomass that characterizes today's areas of upwellings. Without a doubt the most distinctive of the teleosts is *Enchodus*—up to 1 meter long, this predatory "saber-toothed fish" had a prominent pair of curved teeth, which could exceed 5 centimeters in length, at the front of each jaw (fig. 6.9)—but in this "phosphate sea" there were much larger fish, such as *Stratodus*, which had an anguilliform body 4 meters long, and *Saurodon*, 2.5 meters long, with an astonishingly pointed, spear-like lower jawbone.

The marine reptiles from Morocco's phosphates constitute a worldwide reference for

▲ Fig. 6.11. (Preceding pages). The Late Cretaceous, western Morocco. **Above**, a beach scene: Two crocodiles, *Ocepesuchus*, ready themselves to exit the water, while above them two pterosaurs, *Phosphatodraco*, wing their way offshore. On the sand, a theropod dinosaur, *Chenanisaurus*, sniffs around, and in the background a troop of titanosaur sauropods is on the move. **Below**, an underwater scene: A large mosasaur, *Prognathodon*, scouts for prey, and a smaller one, *Carinodens*, feeds on a crustacean at the seabed. Two very strange animals—a plesiosaur with an inordinately long neck, *Zarafasaura*, and a turtle sporting a pipette-shaped snout, *Ocepechelon*—glide along (on the left). *Enchodus*, a teleost (center), *Scapanorhyncus*, a shark (on the right), and *Rhombodus*, a ray (on the bottom), complete the scene.

the Upper Cretaceous/Lower Paleogene interval. They are very abundant, and at least fifty-five species—including squamates, crocodyliforms, plesiosaurs, turtles, and birds—have been unearthed, and the specimens are often complete and spectacular.

During the Maastrichtian the mosasaurs were very diverse, both from a systematic point of view (more than fifteen species have been described) and from an ecological point of view (see fig. 2.80, p. 94). The mosasaurian fauna includes opportunistic mega-predators more than 10 meters long, such as *Mosasaurus* (fig. 6.13), with teeth as sharp as knives, and *Prognathodon* (fig. 6.11), with more massive and robust teeth, which must have fed on bony and shelled prey; it includes medium-sized forms (3–6 meters), such as *Globidens* and *Carinodens* (fig. 6.11), with short and bulbous teeth that must have been used to grind shellfish or crustaceans; and it includes some small forms, such as *Halisaurus*, with a long, slender snout and small pointy teeth for trapping slippery fish.

The plesiosaurs are represented by the elasmosaurid *Zarafasaura*, which sported a very long neck and a very small head. Large chelonioid sea turtles, similar to today's leatherback, included *Ocepechelon* (fig 6.11; also fig. 2.65, p. 84), which sucked up food through its pipette-shaped snout, and *Alienochelys*, with a robust skull and strong jaws for crushing prey. Completing the picture are the lizard *Pachyvaranus*, characterized by a greatly pachyosteosclerotic skeleton (see "The Secrets of Bone," p. 129), and the small gavialoid crocodile *Ocepesuchus* (fig. 6.11), so rare that it is known only from a very fragmentary skull, which raises the possibility that *Ocepesuchus* lived in fresh water and the remains of this individual were carried to the sea. That all these different reptiles lived in the same environment at the same time attests not only to a very precise dividing of ecological niches but also to the significant food base in this "phosphate sea" thanks to the upwellings of its nutrient-rich cold waters.

The Paleogene Mirror

During the Cretaceous/Paleogene crisis, most of these reptiles disappeared, and other ecosystems soon took their place, both in the marine environment and in the continental countryside. This is how a varied fauna of seabirds replaced the pterosaurs. In similar fashion, on the continents, several groups of mammals replaced the non-avian dinosaurs.

In the seas, the mosasaurs and plesiosaurs gave way to a rich and varied collection of crocodyliforms. In the Paleogene, the dyrosaurids in particular (see chapter 2, p. 72, and fig. 2.54), but also crocodiles much like the ones of today, mirrored the mosasaurs of the Maastrichtian, both from the point of view of systematic diversity (there were more than ten species of crocodyliforms) and from an ecological point of view (they could be small or giant, with a long or a short snout, fish eaters or mega-predators, etc.). The lizard *Pachyvaranus* relinquished its place to the sea snake *Palaeophis*. Sea turtles were still very diverse but were now represented essentially by a single family, the bothremydids. The same goes for cartilaginous and bony fishes, which after the Cretaceous/Paleogene crisis were still highly varied but not necessarily represented by the same forms as before.

The Cretaceous/Paleogene Extinction Event

The biological crisis that unfolded at the very end of the Cretaceous (represented in the geologic record by what is often abbreviated as the K/Pg boundary—where K stands for the German word for Cretaceous, Kreidezeit) marked the end of the Mesozoic era and the beginning of the Cenozoic era, but it has been given so much media attention that we shall not go into details here. To briefly summarize the facts, for much of the twentieth century this crisis was ascribed to climatic variations and fluctuations in sea level. Then, in 1980, American geologist Walter Alvarez and his collaborators discovered an unusual spike in iridium—an element from the platinum family that is exceedingly rare on Earth but is frequent in meteorites—in the thin layer of clay that marks the K/Pg transition in Gubbio, Italy. This iridium anomaly was later found in numerous samples from the K/Pg boundary—for example, in El Kef, in Tunisia; in Stevns Klint, in Denmark; and in the Basque Country (Bidart, Zumaia, etc.) (fig. 6.14)—thus pointing to some kind of planetwide event.

The source of this iridium anomaly was soon identified: a giant meteorite that struck Mexico's Yucatan Peninsula near the small town of Chicxulub 66 million years ago. The evidence of this ancient impact is a subterranean crater between 180 and 350 kilometers wide. Buried under Tertiary sediments, it was discovered thanks to seismic measurements. Judging by the dimensions of the crater, the meteorite must have been more than 10 kilometers in diameter. Such a collision would have released about as much energy as a billion of the atomic bombs that were dropped on Hiroshima and Nagasaki! The unimaginable force of that explosion, the ensuing tsunami, the debris and dust flung into the atmosphere, and the long-term effects of this sudden event must have been devastating and could explain the great biological crisis of that time. As further support for some sort of calamity from space, the characteristic minerals we see as a result of meteorite impacts—such as nickel-rich spinels, tektites (drops of molten rock), and shocked quartzes—can be found at the K/Pg transition all over Earth.

But science never rests on claims of having the last word: around the time that the Chicxulub crater was detected, increased iridium levels in the lava of an active volcano in Hawaii were discovered. Iridium is also present in Earth's mantle and crust; thus, it is believed that a certain kind of volcanic activity, known as "**hot spot**" vulcanism, can generate iridium at Earth's surface. This discovery led some scientists, such as French geophysicist Vincent Courtillot, to propose an alternative hypothesis: a violent and protracted bout of vulcanism at the end of the Cretaceous was responsible for the extinctions. In this case as well, incriminating evidence was rapidly found: the Deccan Traps, situated in India's west, are an immense basalt flow that is almost 2,400 meters thick and covers an area somewhat larger than California. These traps date from 68 to 63 million years ago and, therefore, straddle the K/Pg boundary. The immense atmospheric outpouring of ash, as well as the considerable emissions of carbon dioxide (CO_2) and sulfur dioxide (SO_2) this volcanic activity would have generated, must have had a major impact on the climate and led to serious ecological upset.

Following these competing major discoveries, meteorite-impact adherents and

Show Me Your Teeth and I Will Tell You Who You Are!

Today, almost all the ecological niches that marine reptiles occupied during the Mesozoic have been filled by mammals—pinnipeds (seals, sea lions, and walruses), sirenians (dugongs and manatees), and cetaceans (whales, dolphins, and porpoises). Though we are well familiar with these animals, there is still much we do not know about their reproductive and feeding behaviors in the wild, because the marine environment makes those difficult to observe. So, to determine marine mammals' diets, we must supplement direct anatomical and behavioral observation with investigative measures, such as looking for traces of predation on the other animals in their environment and analyzing the stomach contents and the excrement of the species we are studying.

In the case of fossil marine reptiles' diets, we have somewhat less to go on—we mostly have only their dental anatomy, some indirect indicators, and the chemical composition of their teeth. Their stomach contents are very rarely preserved, although we have been able to infer the diet of a couple of species by looking at the remains of prey they ingested. As far as traces of predation are concerned, if any exist (e.g., teeth marks on ammonites or marine vertebrates), they remain anecdotal and usually difficult to match to a species. One simple trace of a bite does not allow one to deduce and then extend this individual instance to general behaviors across the species or groups that might have produced it. Bite marks could just as easily be tied to fights or social behaviors as they could to predation. Last, it is very difficult to connect coprolites (fossilized excrement) to a specific group of organisms.

In view of these difficulties, dental anatomy provides the strongest basis on which to evaluate or propose some hypotheses regarding the dietary preferences of fossil faunas. And, indeed, the shape and size of an animal's teeth give us precious information regarding its diet: from form, one can deduce function (this is one of Georges Cuvier's principles of comparative anatomy). The teeth of Mesozoic marine reptiles exhibit many similarities to those of modern marine mammals, even though the origins of these groups are very different (see "Convergence," pp. 198–99), and we can get some idea of the reptiles' diet by starting from what we observe in the mammals. There are nonetheless two major differences between these groups: (1) mammals have only two generations of teeth (milk teeth and permanent teeth)—unlike reptiles, which constantly renew their teeth throughout their lives—and (2) the great majority of terrestrial mammals exhibit a marked heterodonty (i.e., with teeth of different shapes along the jaw/tooth row). In mammals, however, the return to aquatic life seems to have been accompanied by a loss of differentiation of teeth: the most advanced marine mammals, such as cetaceans, exhibit significant **homodonty**, like that observed in marine reptiles.

In the 1980s, paleontologist Judy Massare, of the University of Rochester, used Mesozoic marine reptiles' tooth shapes to extrapolate characteristics of their diet, based on the known relationships between tooth shape and diet among living marine mammals. She identified seven types of teeth, four of which are shared by today's marine mammals and the Mesozoic marine reptiles and three of which are known only from the latter, based on several criteria: (1) the amount of wear on the tooth (heavy wear characterizes predators at the top of the food chain, capable of capturing and manipulating large prey with bones that can wear down and fracture teeth); (2) the size of the tooth relative to the animal's skull; (3) the shape of the tooth; and (4) the presence and number of cutting surfaces, whether serrated or not. In this manner, Massare was able to define dietary types, or "guilds": groups of predators that shared the same type of prey as a function of their dental morphology. The boundaries between guilds are somewhat arbitrary; there is some crossover, and some teeth are of a type that is between guilds, a fact that reflects the complexity of real-life ecosystems.

These seven guilds can be represented as a triangle, with the three most extreme dental morphologies—piercer, cutter, and crusher—at the corners (fig. 6.12). This can be overlaid with another triangle, of the prey each guild may have preferred: soft prey (fish and invertebrates), bony prey (large vertebrates), and hard prey (ganoid scaly fish, such as sturgeon and gar, and invertebrates with shells) (fig. 6.12). This allows us, for those deposits where the sampling is particularly good, to reconstruct possible food chains and to follow the diversification of certain groups at the expense of others, as well as their evolution.

This method has some limits, in that it cannot determine fossil animals' diets with a high degree of precision, since even today's faunas do not have strict dietary regimens corresponding exactly to their dental characteristics. In fact, a certain portion of predators' diets (including those of today's marine mammals) is opportunistic in nature, as well as dependent on the season and the local environment. For example, although walruses have teeth that are best for crushing, and they eat mostly mollusks and crustaceans, they will not disdain some fish or a lost seal pup. Still, this method allows us to define some dietary preferences of and

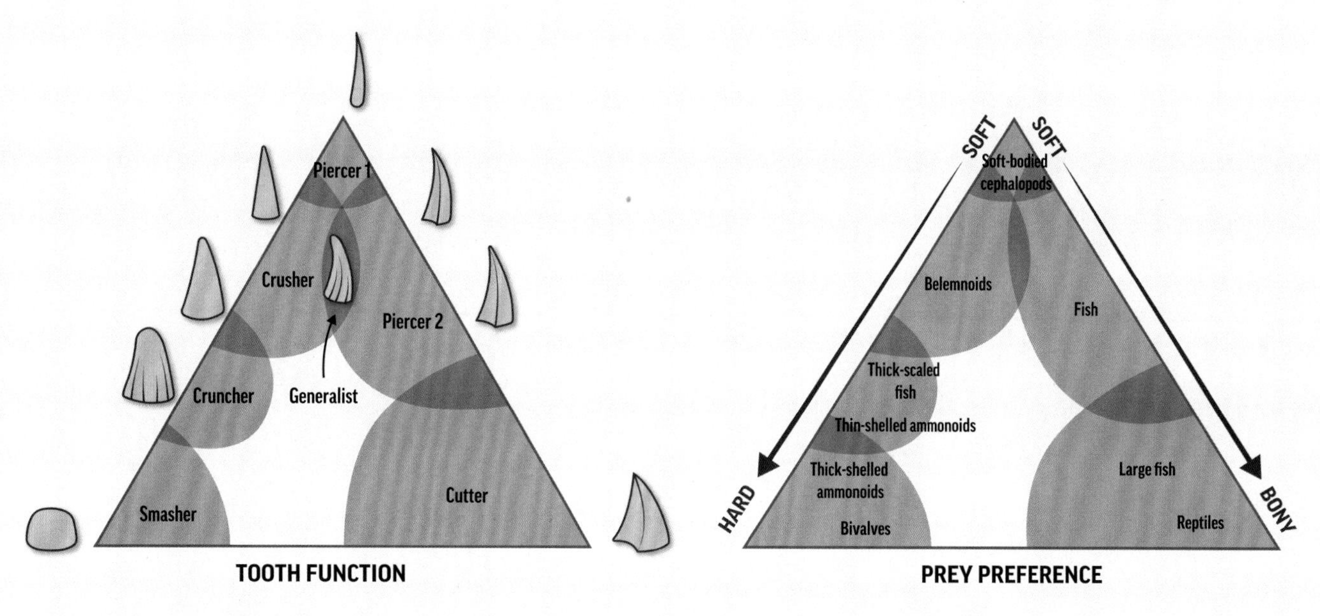

▲ Fig. 6.12. The classification of tooth shapes among predatory marine reptiles (left), leading to extrapolation of their diets (right).

some predation possibilities for extinct marine creatures, thus giving researchers a way to follow trends over time with some consistency.

Additionally, let us not forget that the marine reptiles shared their environment with other large and redoubtable predators, such as sharks and certain bony fish, just as marine mammals do today. Sharks' teeth are quite different from marine reptiles', because of the structure of their jaws.

The anatomical differences notwithstanding, studies similar to those undertaken by Massare and based on marine reptiles (mosasaurs and plesiosaurs) as well as on both cartilaginous (sharks and rays) and bony fish from the Maastrichtian stage in Morocco's phosphates have shown certain convergences that probably indicate relatively close diets. Among the mosasaurs in particular, the three most extreme guilds defined by Massare were confirmed, and it was demonstrated that these guilds were maintained for the entire duration of the Maastrichtian (roughly 6 million years), pointing to fairly constant trophic relations among different groups of animals and a certain stability in the food chain. Finally, in this Maastrichtian series, among the mega-predators we see a tendency to increase in size over time. This tendency was shared by the mosasaurs and the lamniform sharks, with a dominance of the genera *Prognathodon* and *Squalicorax*, respectively.

Geochemical analysis, too, can help researchers better understand the way in which marine communities were structured. Theoretically, the proportion of stable isotopes of calcium in an animal's bones allows us to determine how high that animal was in the food chain. Analysis of the isotopic composition of the calcium in teeth of mosasaurs, plesiosaurs, and large sharks found in the Morocco phosphates reveals proportions that are very similar to each other and quite different from the proportions present in the small sharks and bony fish of the time. These results seem to indicate that mosasaurs, plesiosaurs, and large sharks—the entire range of species—were all at the apex of the food chain, in direct competition with one another. This relatively rare configuration was enabled by some very particular environmental conditions, where a biomass of not very differentiated prey was the food source for the entire set of predators in this famous area of upwellings (see p. 172). These results that at first seem to contradict the anatomical data are a precious complementary element, demonstrating how important it is to take into account the environment in which all these animals lived so as to best discern the trophic structuring of these fossil communities in all their dimensions.

Top Predator: *Mosasaurus*

Name: "lizard of the Mosa"
Classification: Squamata, Mosasauroidea, Mosasauridae
Lived: Late Cretaceous (Santonian–Maastrichtian, 83–66 Ma)
Range: worldwide
Skull length: more than 1 meter
Overall length: up to 15 meters
Diet: fish, cephalopods, other marine reptiles

Mosasaurus was, together with the Triassic ichthyosaur *Shonisaurus*, the largest marine reptile of the entire Mesozoic. But whereas the ichthyosaur was probably a placid giant like today's baleen whales, *Mosasaurus* was a redoubtable mega-predator to which every creature was potential prey. Its powerful jaws were armed with teeth that were both robust and sharp as knives. Any creature unlucky enough to be caught between those jaws but somehow still

alive was prevented from escape by the backward-pointing teeth lining the roof of *Mosasaurus*'s mouth.

Its very long, hydrodynamic body; its powerful swimming paddles, which were among the most specialized of all mosasaurs'; its global distribution; and its twelve species all show that *Mosasaurus* was probably a great hunter that crisscrossed the seas, as orcas do today. *M. hoffmanni*, the type species of the genus, was the first mosasaur to be discovered, at the end of the eighteenth century in the Netherlands. That fossil left scholars of the time perplexed, until Georges Cuvier (see "Georges Cuvier," p. 95) determined that this was a reptile similar to a monitor lizard ... only a giant, extinct marine one.

▼ Fig. 6.13. *Mosasaurus*, a mosasauroid from the Upper Cretaceous around the world.

► Fig. 6.14. The Cretaceous/Paleogene (K/Pg) boundary in Zumaia, Basque Country. The layer that marks the K/Pg boundary (indicated by the arrow) is clearly visible between the gray marls of the Maastrichtian stage on the right and the pink limestones from the Danian on the left.

catastrophic-vulcanism adherents argued the reason for the K/Pg mass extinction for decades. Today the vulcanism hypothesis has lost ground because several works have shown that it could not, by itself, have led to the worldwide wipeouts that occurred. In addition, (a) the quantity of iridium erupted by volcanoes remains modest and does not provide an explanation for its abnormal concentration at the K/Pg boundary; (b) the ratio of iridium to palladium (another element from the platinum family) at the K/Pg boundary is very different from the ratio found in the lavas of the Deccan Traps (and close to that found in meteorites); and (c) the iridium peak was highest in the middle of the traps series, demonstrating that the volcanic effusion started well before the extinction of 66 million years ago. Yet, if vulcanism was not the principal cause of the extinction, it may have contributed—just as the great marine transgression that characterized this crisis did—to the devastating effect of the meteor's impact. All together, these disturbances must have projected an immense quantity of dust, ashes, and toxic gases into the atmosphere, leading to a darkening of the atmosphere for months or even years, thus disturbing both photosynthesis and the whole range of ecosystems. The result would have been a general breakdown of the food chains.

Once we ponder the message the fossils send us today, however, things are not so simple. Let us point to the most important caveat: plants, which turn sunlight into energy and are consumed by herbivores, were only very slightly affected by the crisis. Yet extinctions in the animal kingdom were significant: 80% of marine microorganisms (foraminifera, nanofossils, etc.) were lost, and ammonites, belemnites, and rudists (bivalve mollusks responsible for immense calcareous bioconstructions, especially during the Cretaceous) disappeared forever. Among the vertebrates, the bony fish and the sharks suffered severely too, and the non-avian dinosaurs and most marine reptiles were wiped out.

To go into detail as concerns the marine reptiles, the mosasaurs and the plesiosaurs went extinct in a rather brutal fashion, because they were diverse and widespread right up to the end of the Cretaceous. That is to say, these two groups were flourishing, with no indication that they might have no longer been sufficiently adapted to their circumstances. Meanwhile, the sea turtles (chelonioids and bothremydids), as well as the dyrosaurid crocodyliforms, were scarcely affected by the crisis, and they diversified particularly well starting in the Paleocene epoch (66–56 Ma) (see chapter 2, p. 85). It is possible the reduction of oceanic plankton (these tiny animals were severely affected, as described above) affected the entire food chain that depended on it. Yet certain large animals, such as numerous crocodyliforms and turtles, in continental environments (but also mammals and birds) depended mostly on freshwater food chains; thus they were less affected by the crisis and, on the whole, survived (see chapter 7, p. 189). This hypothesis seems to be corroborated by studies of deep-sea fish that relied on dead matter (organic debris) sinking down from the upper waters. Such fish do not and did not depend directly on photosynthetic activity, and the Cretaceous/Paleogene crisis worked out very well for them! Differences in physiology among marine reptiles, too, could help explain why only some groups survived. Maintaining an internal body temperature higher than the surrounding temperature requires a great deal of energy. During an intense but relatively short crisis, ectotherms (cold-blooded animals), such as crocodyliforms and turtles, which consume much less energy, can wait for the proverbial storm to pass. The large, energy-devouring endotherms (see "They Were Warm-Blooded!," pp. 154–55) would have had no such luxury. Whatever the case may be, decades of all-around research and hundreds of scientific articles published in fields as varied as paleontology, geophysics, and climatology have still not allowed us, on the one hand, to satisfactorily explain the causes of this crisis and, on the other, to cast some light on the reasons for its different impacts on different groups of organisms. As in the case of the other crises, probably a combination of factors was responsible.

CHAPTER 7

THE CENOZOIC ERA

Illustration by Augustus Burnham Shute for *Moby-Dick*, Herman Melville, 1851.

I

What's Left of the Cretaceous?

After the Cretaceous/Paleogene crisis—with no mosasaur or plesiosaur left alive, and with the ichthyosaurs having disappeared at the beginning of the Late Cretaceous—the marine landscape was profoundly different. The sharks and the rays, particularly the epipelagic and mesopelagic forms (those that lived in the open ocean, at depths of less than one thousand meters), even though they did not disappear completely, had perished in great numbers too. The oceans therefore were deprived of all the most redoubtable predators of the Mesozoic.

Nonetheless, the sharks bounced back very quickly, reestablishing their hegemony and their place at the apex of the food chains by the mid-Paleocene. But it was a group of reptiles that managed to really take advantage of the crisis and colonize the marine environments that the mosasaurs had vacated. In fact, not all marine reptiles had been affected by the crisis in the same manner: while some large taxonomic groups went extinct, other groups survived despite a loss of many species, and still other groups were not harmed in the slightest and even benefited from the crisis. One group's loss was the other's gain, and it seems that the dyrosaurids—a lineage of crocodyliforms we have already mentioned (see chapter 2, p. 66)—seized their cousins' disappearance as an opportunity to diversify, to spread, and to occupy new ecological niches. For them, the Cretaceous/Paleogene crisis was a time of plenty. Like the dyrosaurids, the crocodilians—the group of crocodyliforms to which all of today's crocodiles belong—had survived into the Paleogene without much problem. Some species of crocodilians even seem to have emerged unchanged. Less seafaring than the dyrosaurids, their way of taking advantage of the situation was to colonize new continents, proliferating in fresh water especially.

Not much more than 10 million years into the Paleogene, the dyrosaurids dominated the marine environments, especially in the southern hemisphere. Whereas during the Maastrichtian they had been restricted principally to African marine basins, in the Paleocene they enjoyed both a diversification and a dispersal that were nothing short of extraordinary. Their success left the crocodilians with only a very small place in the oceans, essentially in the northern hemisphere, where the dyrosaurids were considerably less diverse. Our knowledge of what was happening in those northern seas is rather limited, because Paleocene marine deposits in those parts are very meager. It may be that the cooler climate compared to that of the southern hemisphere limited the spread of the dyrosaurids, which seem to have preferred more tropical temperatures.

A question arises nonetheless. Why did the crocodyliforms not share the fate of the other predatory reptiles, both in the seas and on land? The first possible answer is that the Cretaceous dyrosaurids and, even more certainly, crocodilians possessed a physiology that was close to that of today's crocodiles. They were, therefore, poikilotherms (incapable of regulating their own body temperature), a physiology that, although it leaves such animals unable to engage in prolonged intense activity without overheating, is particularly economical from an energy-use standpoint.

Given they are the same weight, a crocodile needs seven times less food than a lion does—which, in times of scarcity, can confer quite an advantage! In addition, the crocodile's slow metabolism can be slowed even further, as when some modern crocodiles fast for months or even in excess of a year. Did this ability allow the crocodiles as a whole (dyrosaurids and crocodilians) to let the severe disruptions in the trophic chain—and the accompanying shortage of prey—pass and to wait for better days, while the large marine reptiles with a more active metabolism could not do without food?

The second possible answer is that, like their living counterparts, the ancient crocodiles spent their early life in fresh water and migrated to marine environments as adults. Some research shows that the Cretaceous dyrosaurids may well have been much less marine than was thought and occupied freshwater environments for the most part. It may be that these environments were less disturbed by the crisis, so the crocodiles in general survived, and the dyrosaurids then colonized the marine environments left vacant.

Unfortunately for the dyrosaurids, they did not profit for very long from their new status as leaders of the marine world: during the Bartonian age of the Eocene epoch, about 40 million years ago, adverse events—a decrease in sea level, a drop in temperatures, and the rise of new competition, including the cetaceans that dominate the oceans today—conspired against them. Since dyrosaurids preferred tropical and coastal habitats, the decrease in warm climates and the draining of continental shelves began to squeeze these reptiles out of existence. The mammals, since their physiology was much more dynamic than that of the crocodilians, did not have the same problems. Did marine mammals put competitive pressure on the dyrosaurids, or did they simply take advantage of their setbacks? That is difficult to establish. Probably the truth is that they did a little bit of both. The dyrosaurids soon perished, leaving no descendants.

Unlike the dyrosaurids, however, the crocodilians tolerated cooler temperatures well. It was only a drastic decrease in the size of their favorite climatic zone that, much later, would reduce their range. In addition, their more coastal and freshwater lifestyle may help explain why they were much less affected by the aforementioned events of the Bartonian age than were the dyrosaurids, which were much more marine and therefore subject to marine regressions and novel competition in this environment.

To return now to the dawn of the Paleocene, the sea turtles also do not seem to have been greatly affected by the Cretaceous/Paleogene crisis, and the number of genera of the two large groups present during the Maastrichtian—the chelonioids and the bothremydids—actually rose during the Paleocene! This would seem to demonstrate that sea turtles, like dyrosaurids, availed themselves of newly empty ecological niches. Why did they prosper so much? For the time being, research is wanting, and the mystery remains. It is nevertheless interesting to note that following the Campanian—their golden age—sea turtles suffered a sizable loss of species in the Maastrichtian but did not experience a comparable loss during the Cretaceous/Paleogene crisis. The setback that caused their reduction after the Campanian remains unknown as well. In any case, in the Paleocene, sea turtles continued their adventure in their element. The bothremydids were particularly diverse during the early Paleogene but also fell victim to extinction during the Lutetian age (ca. 45 Ma). The chelonioids were quite diverse throughout the Paleogene and the Neogene periods, and two of the three Cretaceous families, the cheloniids and the dermochelyids, survive today (see chapters 1 and 2, pp. 9 and 82). Only the family of the protostegids, which included species of large—indeed, very large—size (see chapter 2, p. 81), vanished alongside the plesiosaurs and mosasaurs 66 million years ago.

Of Dinosaurs: To Fly or to Swim? Why Choose?

▼ Fig. 7.1. *Hesperornis* and *Ichthyornis*, two marine birds from the Upper Cretaceous in the United States, and the skull of *Hesperornis*, showing the teeth. As a matter more of curiosity than of importance, the fossilized jaws of these avians were sometimes mistaken for those of mosasaurs (because of their teeth).

Although most Mesozoic dinosaurs were terrestrial, some marine species did exist ... represented by birds! There were, therefore, "marine dinosaurs," but not in forms that earlier paleontologists would recognize as reptilian (see "Reptiles? Sauropsids? What's the Fuss About?," p. 27). As we already stated (see p. 29), no non-avian dinosaur is considered to have been marine. A review of the marine birds and their evolutionary history would be so lengthy as to require a second book, but we cannot resist providing a very basic overview of this unique bestiary that navigated between the sea and the sky.

Starting in the Late Cretaceous, different groups of birds adapted to the aquatic environment. The ichthyornithiforms and the hesperornithiforms are known from the deposits of the Western Interior Seaway in North America, with defining species including *Ichthyornis* and *Hesperornis*, marine

◄ Fig. 7.2. A size comparison of several modern penguin species and *Icadyptes*, a giant penguin from the Eocene in Peru.

birds with small pointy teeth (fig. 7.1). These groups did not survive the Cretaceous/Paleogene crisis and passed the baton to a new group, the neornithes. Some remarkable neornithes, the penguins (fig. 7.2)—one of the groups of birds with the most significant adaptations to the marine environment—first appeared in the early Paleocene, about 65 million years ago. Their rapid rise, so soon after the crisis, demonstrates the group's evolutionary dynamism and noteworthy diversification.

III

Marine Mammals

Although mammals populate the fossil record starting in the Triassic, there wasn't any place for them in the Mesozoic seas, which were saturated—as far as tetrapods are concerned and as we have covered in the bulk of this work—by reptiles. Today, mammals have taken their place at the apex of the food chains. But they were slow to do so, even following the Cretaceous/Paleogene crisis. In fact, mammals conquered the oceans much later than the birds did, starting in the Eocene, almost 15 million years after the mass extinction. The reason is because of the other reptiles we just discussed, which themselves had awaited their turn to shine: the crocodyliforms. The mammals succeeded in this arena only once the marine crocodyliforms' habitat began to contract.

Today's marine mammals include both semiaquatic and exclusively aquatic forms. The first are representatives of the order Carnivora: the pinnipeds (sea lions, seals, and walruses), the sea otter, and the polar bear. The second are cetaceans (whales, dolphins, and porpoises) and sirenians (manatees and dugongs). Starting in the Eocene, the fossil record testifies to a great diversity of marine mammals—including some failed lineages, such as the strange desmostylians, from the early Oligocene to late Miocene on the north side of the Pacific; the aquatic sloth *Thalassocnus*, from the late Miocene to late Pliocene (ca. 9 to 5 Ma) in South America (fig. 7.3); and the aquatic bear *Kolponomos*,

▲ Fig. 7.3. *Thalassocnus*, an aquatic sloth from the Miocene and the Pliocene in South America.

from the early Miocene (ca. 20 Ma) in the North Pacific. The sea otters, for their part, first appeared during the late Miocene (ca. 10 Ma); the polar bears, in the late Pleistocene (ca. 100,000 years ago).

Because of their ectothermy and, hence, difficulty maintaining a sufficient body temperature in cold waters, today's marine reptiles are restricted to tropical or subtropical regions, except for the leatherback turtle, *Dermochelys*, which, thanks to its large size and good insulation, can swim in the coldest waters. By contrast, the marine mammals, thanks to their endothermy, occupy all latitudes and in some cases travel from one pole to the other.

The Cetaceans

The cetaceans (together with the sirenians, discussed next) were the first mammals to make their appearance on the marine stage, during the early Eocene, about 50 million years ago. The most ancient fossils have been found in the region of the India-Pakistan border, which at the time was part of the Tethys between India and Asia (see p. 13). These primitive whales, the archaeocetes, comprised five families: Pakicetidae, Ambulocetidae, Remingtonocetidae, Protocetidae, and Basilosauridae.

The pakicetids (about the size of a wolf), known only from Lower and Middle Eocene deposits 55 to 38 million years old, are considered the most primitive cetaceans (fig. 7.4a). Although they had not undergone much overall modification for aquatic life, certain features of their skeletons (e.g., osteosclerosis of the limbs and the ribs, eyes close together and toward the top of the skull) tell us that they must have walked or swam on the bottom in shallow aquatic environments.

A greater number of characteristics associated with an aquatic mode of life can be observed in the ambulocetids (fig. 7.4b)—which had a large and long skull, a long and powerful tail, and limbs that were long but ended in paddles—and in the remingtonocetids (fig. 7.4c), with their long body and shorter limbs, which lived in the mid-Eocene, between 48 and 38 million years ago. At that time, the protocetids (fig. 7.4d) were the first cetaceans to spread through the oceans. We know this because their fossils are found not only in India and Pakistan, but also in Africa, Europe, and North America. Some are characterized by a lack of connection between the pelvic girdle and the spine (as in later mosasauroids), suggesting an exclusively aquatic way of life.

By the mid- to late Eocene, around 38 million years ago, the basilosaurids (fig. 7.4e) already rather resembled today's whales, with their forelimbs transformed into swimming

paddles, their hind limbs almost nonexistent, and their spine and tail adapted to an oscillating mode of swimming and probably ending in a horizontal caudal fin, for powerful propulsion. At the end of the Eocene epoch and, even more so, during the late Oligocene (roughly 25 Ma), the archaeocetes suffered a reduction as the neocetes, representing today's whales, diversified. The neocetes are divided into two groups: the mysticetes, or baleen whales (right whales, fin whales, etc.), and the odontocetes, those with teeth (dolphins, orcas, sperm whales, etc.). Their vast variety in size notwithstanding, modern cetaceans have a very consistent overall appearance, with a long and tapered body, a short neck, swimming paddles at the front, an absence of hind limbs (there is only the remainder of a pelvis), and a powerful horizontal tail.

▲ Fig. 7.4. a. Pakicetidae (*Pakicetus*); b. Ambulocetidae (*Ambulocetus*); c. Remingtonocetidae (*Kutchicetus*); d. Protocetidae (*Maiacetus*); Basilosauridae (*Darudon*).

Top Predator: *Basilosaurus*

Name: "king lizard"
Classification: Cetacea, Archaeoceti, Basilosauridae
Lived: Mid- to late Eocene (Bartonian–Priabonian, 40–36 Ma)
Known range: southern United States (Louisiana, Alabama, Mississippi, etc.), Egypt, Pakistan
Skull length: 1–2 meters (fairly short in proportion to the animal's size)
Overall length: up to 21 meters
Diet: fish, cetaceans

Basilosaurus, the first archaeocete (a kind of primitive cetacean) to be described, was the largest cetacean of the Eocene epoch. At the time of its discovery in 1834, American paleontologist Richard Harlan thought it was a sea snake and classified it among the reptiles—hence its name ending in *saurus*, meaning "lizard." In 1839, English

anatomist Richard Owen determined it was a mammal and renamed it *Zeuglodon*, meaning "yoke-tooth," in reference to its distinctive tooth shape. Nevertheless, the nomenclatural code to which all biologists and paleontologists refer stipulates that the first published name must be kept, even if it was based on an erroneous conclusion.

All three species of *Basilosaurus* (*B. cetoides*, *B. isis*, and *B. drazindai*) possessed a very long body, forelimbs in the shape of swimming paddles, and very small hind limbs. Since the hind limbs were not connected to the spine, they cannot have been of any use in swimming, but they may have been used to facilitate mating.

▼ Fig. 7.5. *Basilosaurus*, a primitive cetacean from the Eocene in the United States, Egypt, and Pakistan.

► Fig. 7.6. *Prorastomus* (above), a sirenian from the Lower Eocene in Jamaica, and today's manatee, *Trichechus* (below).

The Sirenians

Today the sirenians, or "sea cows," include two families: Trichechidae (manatees) and Dugongidae (dugongs). They both have a thickset body, a horizontal caudal fin, some very short forelimbs transformed into swimming paddles, and no hind limbs. Strictly herbivorous, they graze vegetation at the bottom of or floating atop shallow aquatic environments (i.e., coastal marine habitats, rivers, and estuaries). Sirenians first appeared during the early Eocene, about 50 million years ago (around the same time as the cetaceans), with the prorastomids *Prorastomus* (fig. 7.6) and *Pezosiren*, fossils of which have been found in Jamaica. Like the first cetaceans, the first sirenians are considered to have been semiaquatic, inhabiting rivers and estuaries. Although most sirenians lived or live in tropical or subtropical waters, Steller's sea cow (*Hydrodamalis gigas*), which reached an incredible 10 meters long, lived in the arctic waters of the Bering Sea up to the eighteenth century before being hunted to extinction shortly after its discovery.

◄ Fig. 7.7. The pinnipedomorph *Enaliarctos*, from the end of the Oligocene and beginning of the Miocene in California and Oregon (United States).

▼ Fig. 7.8. Two desmostylians: *Ashoroa*, from the Upper Oligocene in Japan, and *Desmostylus*, from the Upper Oligocene to the Upper Miocene in the north coast of the Pacific.

The Pinnipeds

Modern pinnipeds include the families Otariidae (the eared seals and sea lions), Odobenidae (the walruses), and Phocidae (the seals). Both their forelimbs and their hind limbs have been modified into swimming paddles but are used for movement on land, albeit in a less agile manner, as well. The pinnipeds (the pinnipedomorphs; fig. 7.7) first appeared in the late Oligocene (about 25 Ma) in the east of the North Pacific (Oregon). The most primitive forms looked similar to today's otters, with short tails and webbed feet.

The Desmostylians

Desmostylia is an extinct order of herbivorous mammals closely related to the sirenians, from the Lower Oligocene to the Upper Miocene in Japan and on the west coast of Mexico and the United States (fig. 7.8). Judging by their general features and the microanatomy of their bones, desmostylians were essentially aquatic. Their fossils have been found in coastal, lagoonal, and estuarine deposits. However, the ecology of these creatures is a mystery. On the basis of various reconstructions, their lifestyle may have resembled that of polar bears, hippopotamuses, sirenians, or even pinnipeds.

Convergence

► Fig. 7.9. A mosasaur and a thalattosuchian crocodyliform compared with an orca.

While evolution is subject to the vagaries of mutations and variations, it is nevertheless framed by the constraints imposed by the environment. These constraints, like those of the aquatic environment, apply equally and generally to all organisms that live in that same environment, therefore sometimes producing animals that resemble one another, even though they have no kinship ties. This is what is called convergence.

It is obvious that—just as for eating soup, a spoon is much more practical and effective than a fork—for swimming, a paddle is more effective than a limb with fingers. Similarly, a hydrodynamic body, tapered and smooth, will move faster in the water than a stocky body with hairs or large scales. Obviously, there are "intermediate" solutions! And this is precisely where things get interesting, because we can use our knowledge of the morphology and ecology of living animals to deduce precious information about the ecology of fossil animals, of which conservation is most often incomplete.

Direct anatomical study of fossil animals' general morphology (their general shape, swimming paddles, auxiliary fins, dental morphology, etc.), combined with histological and geochemical studies that tell us about their metabolism, already allows us to determine a great deal about their presumed way of life. But by looking at these extinct creatures through the lens of our understanding of modern animals, we can go even further. For example, it has often been suggested that pliosaurs, mosasaurs, and even some thalattosuchian crocodyliforms must have occupied ecological niches similar to those of today's marine mega-predators, the orcas (fig. 7.9). Likewise, it is supposed that the large shastasaurid ichthyosaurs of the Triassic occupied ecological niches comparable to those of baleen whales and beaked whales (which feed using suction and aspiration, as the shastasaurids likely did) (fig. 7.10). This supposition may be accurate even though no marine reptile was equipped with baleen.

Most ichthyosaurs and dolphins also exhibit obvious similarities (fig. 7.11), which they share with some sharks and bony fish, such as tuna and swordfish:

▲ Fig. 7.11. An ichthyosaur compared with a dolphin.

a hydrodynamic body with swimming paddles in the front, hind limbs (pelvic fins, in sharks and bony fish) that are either small or absent, a dorsal fin, and a bilobed caudal fin, with a narrow point of attachment.

The evolution of limbs into swimming paddles took millions of years. The most primitive aquatic tetrapods still possessed supporting limbs, while the most advanced ones were equipped with swimming paddles, useless for movement on land. In the case of "intermediate" forms, however, it is

▲ Fig. 7.10. A shastasaurid ichthyosaur compared with a baleen whale.

▲ Fig. 7.12. A nothosaur compared with a mosasaur and a primitive cetacean.

▲ Fig. 7.13. A placodont compared with a desmostylian and a sirenian.

often difficult to distinguish those that might have been completely independent of the terrestrial environment from those that must have occasionally returned to dry land. For example, it is difficult to determine whether the nothosaurs, the primitive mosasaurs, and the primitive cetaceans (pakicetids and ambulocetids) (fig. 7.12) had become exclusively aquatic or were still capable, at least occasionally, of terrestrial locomotion. One peculiarity of the lineage to which cetaceans and sirenians belong is the complete loss of the hind limbs, whereas all marine reptiles (with the exception of most sea snakes) retained them.

Although greatly differing in origin, some marine creatures—such as today's sirenians, the Paleogene and Neogene desmostylians, and even the Triassic placodonts (fig. 7.13)—exhibit a stocky body and an increase in bone mass, characteristic of animals that graze at the bottom of shallow aquatic environments.

On the other hand, some morphologies and ecologies are one of a kind, with no past or present equivalent. In the fossil record, we often find animals with bizarre characteristics that even paleontologists could not have dreamed up, such as the ridiculously long neck of the elasmosaurid plesiosaurs or of the archosauromorph *Tanystropheus* (fig. 7.14). On the opposite end of the time scale, some of today's animals, such as seals and sea lions, have no counterpart among the marine reptiles of the Mesozoic (fig. 7.14).

Convergences are not limited to similarities between fossil forms and current ones. We find analogies within the ranks of the Mesozoic marine reptiles too; for example, certain thalattosuchians (e.g., *Dakosaurus*; fig. 5.21, p. 161) and large mosasaurids (e.g., *Mosasaurus*; fig. 6.13, pp. 182–83) display notably comparable dentition, adapted to a very specific way of life: opportunistic mega-predation.

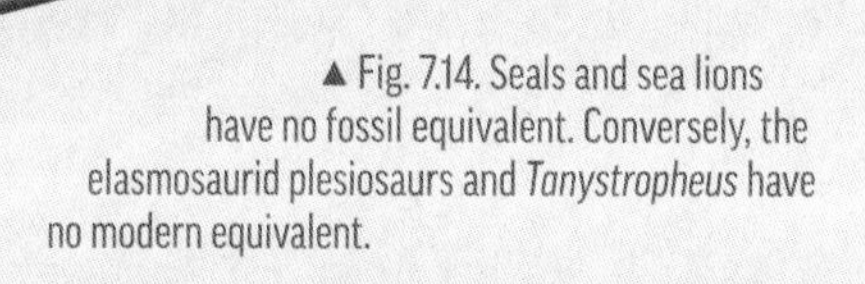

▲ Fig. 7.14. Seals and sea lions have no fossil equivalent. Conversely, the elasmosaurid plesiosaurs and *Tanystropheus* have no modern equivalent.

GLOSSARY

actinopterygians: saltwater or freshwater fish that are ray-finned; most of today's fish belong to this group.

amniotes: tetrapod vertebrates (reptiles [including birds] and mammals) equipped with an amnion, a solid envelope protecting the embryo (which is developing in the amniotic fluid) from shocks and dehydration.

analogy: an anatomical structure in one species that exhibits resemblances to a structure in another species, tied not to kinship relations but to adaptations acquired via convergence due to pressures from similar environments. The wings of a bird, the wings of a bat, and the wings of a pterosaur are analogous.

anoxia: lack of oxygen in the marine environment. Anoxia favors an often exceptional preservation of fossils, something that translates visually to very dark sediments, rich in organic matter.

archosauromorphs: a clade of diapsid reptiles that includes crocodiles, birds, and numerous fossil groups, such as the non-avian dinosaurs and the pterosaurs.

autopods: the distal (lower) segment of the limbs of tetrapod vertebrates, corresponding to the hands and feet. Each autopod includes the carpals or tarsals (the wrist or ankle bones), the metacarpus or metatarsus (the structure that forms the palm or instep), and the phalanges (the finger or toe bones).

benthic: descriptive of sessile or mobile organisms that live on the bottom of bodies of water (oceans, lakes, rivers).

bioturbation: an alteration of the first few centimeters of soil or sediments beneath the surface due to organisms living in these environments.

biozone: a stratigraphic unit defined by an association of fossil **taxa** (such as certain species of ammonoids) and that can be used to classify and date geological strata.

brachiopods: very ancient marine animals equipped with two valves (a ventral one and a dorsal). They are often confused with bivalve mollusks, which are also endowed with two valves, but those valves are oriented differently (a left and a right).

Chondrichthyes: fish with a cartilaginous skeleton, including the elasmobranchs (sharks and rays) and the holocephalans (chimaeras).

cladistics: a method of classification with the goal of identifying kinship relations between organisms, represented in the form of a "tree" called a cladogram. The proximity between "branches" of the tree indicates the degree of kinship.

connective tissue: animal tissue made of fibers, with a function of supporting and protecting the organs.

continental shelf: a part of a continent situated below the ocean's surface.

convergence: a characteristic (morphological, physiological, or behavioral) that is similar between two taxa that are not strictly related and which, therefore, rather than having been acquired via a common ancestor, signifies two separate but similar responses to an environmental constraint.

diagenesis: the set of physical, chemical, and biological processes that transform sediment into rock and that may affect the fossilization of organic remains.

diapause: a temporary halt in the development of an organism, during which metabolic activity diminishes.

diapsids: reptiles—including birds—in which the skull has (or used to have) two temporal (on each side of the head) fossae (or windows), openings where the muscles that control movement of the jaws are inserted.

durophage: an animal that feeds on hard prey (organisms with shells or exoskeletons, corals, etc.) that must be crushed.

ectotherm: an animal in which body temperature is dependent on the temperature of the environment. Classically called "cold-blooded."

endotherm: an animal that regulates its own body temperature. "Classically called "warm-blooded.

epicontinental sea: a sea covering a part of a continent. Such bodies of water rarely exceed 200 meters in depth.

evolutionary radiation: an explosion of diversity, often following a mass extinction. One speaks of "adaptive radiation" once the niches that were left empty are conquered by a lineage.

filter feeder: an animal that feeds by filtering out minuscule organic particles (i.e., plankton) or inorganic particles suspended in the water.

gigantothermy: the capacity to maintain an elevated body temperature thanks to a large body size and good thermal insulation.

haernal spine or **chevron**: a bony blade or point situated below the main body of the caudal vertebrae in certain vertebrates.

heterodont: an animal equipped with several types of teeth, with different forms and functions. Most mammals are heterodonts.

homodont: an animal that is equipped with a single kind of tooth. Most reptiles are homodonts.

homology: an anatomical structure that is the result of a kinship relation between distinct organisms and is inherited from a common ancestor. The limbs of tetrapod vertebrates, composed of the same bones—even if they are specialized into a wing or a swimming paddle—are homologous.

hot spot: a very localized area situated inside a tectonic plate where magma from the mantle rises to the surface by means of regular volcanic activity.

hydrofoil: from the boating world, referring to a streamlined wing that facilitates movement, speed, and stability in water. In an extended usage of the term, it is used to qualify the swimming paddles of aquatic vertebrates.

hypocercal: refers to a caudal fin in which the larger lobe, the one that includes the final bones of the tail, is oriented downward. This is the case, for instance, in ichthyosaurs and thalattosuchian crocodyliforms.

keratin: a protein that is the principal component of skin appendages, such as hair, feathers, horns, scales, and nails.

kinesis: when the bones of the skull and the mandible are not fused but slide on and between each other, allowing for the ingestion of large prey. This mechanism, which is pushed to extremes in snakes, also existed among the mosasaurs.

Lagerstätte: this is a reference to the German mining term, which means something like a "storage area"; a sedimentary deposit containing fossils that are more numerous, more diversified, and better-preserved than usual. Konzentrat-Lagerstätten are deposits with very large numbers of fossils but fossils that are disarticulated, and Konservat-Lagerstätten are deposits with fossils that are very well preserved.

lepidosauromorphs: a clade of diapsid reptiles that are the sister-group to the archosauromorphs and that include the sphenodontids, the squamates (lizards, snakes, and amphisbaena), and numerous other fossil groups.

marine regression: a drop in sea level leading to an increase in land area and, conversely, to a shrinking of epicontinental seas.

marine transgression: a rise in sea level that leads to a decrease in land area and, conversely, to an expanding of epicontinental seas.

medullar cavity: the central, often cylindrical, cavity in long bones.

monophyletic: describes a group of organisms consisting of the set of organisms that are descended from a single common ancestor and that are defined by the shared characteristics inherited from this ancestor.

neural spine: the bony blade or point situated below the main body of all vertebrae.

pelagic: describes an organism that lives in the open ocean, far from the coasts and above the ocean floor.

oceanic ridge: a chain of underwater mountains that form around a rift, an area of divergence between tectonic plates and from which magma originating in Earth's mantle pours out.

oceanization: the process of an ocean's formation, following continental fracturing. Today this phenomenon is observable in the African rift, which runs from north to south: in several million years, an ocean will separate the land to the east of the rift from the land to the west.

ontogenetic stages: the different developmental stages of an organism, from its conception to its maturity, including its death.

Osteichthyes: fish with a bony skeleton, which include the **actinopterygians** and the **sarcopterygians**.

osteosclerosis: an increase in bone density, corresponding to an adaptation that is generally observed in coastal marine animals that inhabit shallow waters.

oviparous: egg-laying.

pachyostosis: a non-pathological increase in bone deposits, observable in bones that have a bloated appearance. This adaptation is generally used by coastal marine animals to increase body mass and reduce buoyancy, allowing them to stay near the shallow seabed with minimal effort.

pachyosteosclerosis: a combination of **pachyostosis** and **osteosclerosis**, resulting in bones that are both denser and more bloated.

paraphyletic: characterizes a group that includes only some of the descendants of a common ancestor.

phylogeny: the study of kinship relations between organisms.

polyphyletic: describes a group that contains a certain number of species but excludes the most recent common ancestor.

sarcopterygians: fish with fleshy fins. They are divided into Actinistia (coelacanths) and Rhipidistia, which include the lungfish and the **tetrapods**.

sclerotic ring: a ring formed by small interconnected bony plates, inserted in the eyeball of certain vertebrates, providing support and protection for the eye.

streptostyly: in squamates, the mobility of the quadrate bone (a bone in the skull that is articulated with the mandible) compared to the rest of the skull, made possible by the loss of the lower temporal fossa.

stylopods: the proximal (upper) segment of tetrapods' limbs, corresponding to the humerus (in each forelimb) and the femur (in each hind limb).

subduction: the convergence of two tectonic plates, where one dives beneath the other.

taphonomy: the branch of paleontology that deals with the processes of fossilization.

taxon (pl. **taxa**): a nomenclatural unit that groups together all the organisms of the same rank and is defined by a set of shared characteristics assembled in a diagnosis. From most similar to least similar characteristics, some taxa are species, genus, family, order, class, phylum, and kingdom.

teleosts: bony fish that are ray-finned, corresponding to almost the entirety of today's fish species and to roughly half of all vertebrate species.

tetrapods: vertebrates equipped with two pairs of limbs, even if some forms (e.g., snakes) have lost them. They include amphibians, reptiles (including birds), and mammals.

thecodonts: a group of reptiles from the Permian and Triassic that is today considered paraphyletic. The word, which means "socket-tooth," also refers to a manner of dental implantation.

traps: vast layers of basaltic flows, produced by intense vulcanism over a very long time. Some, like those in Decca (India) and Siberia, coincide with mass extinctions in the fossil record.

trilobites: very diverse marine arthropods from the Paleozoic era, with a body composed of three longitudinal segments (hence the name). Trilobites became extinct during the Permian/Triassic crisis.

upwelling: occurs when cold ocean waters very rich in plankton rise toward the surface. A great diversity and quantity of organisms, notably fish, are attracted by this food.

viviparous: describes animals that give birth to live young rather than lay eggs.

zeugopods: the midsection of tetrapods' limbs, corresponding to the forearm (radius and ulna) and the lower leg (tibia and fibula).

ziphodont: describes teeth with serrated carinas. Such teeth are generally long, blade-shaped, and curved. They work like knives and are characteristic of carnivorous animals.

BIBLIOGRAPHY

Allemand, R., N. Bardet, A. Houssaye, and P. Vincent. "Endocranial Anatomy of Plesiosaurians (Reptilia, Plesiosauria) from the Late Cretaceous (Turonian) of Goulmima (Southern Morocco)." *Journal of Vertebrate Paleontology* 39, no. 2 (2019): e1595636.

Bardet, N., J. Falconnet, V. Fischer, A. Houssaye, S. Jouve, X. Pereda Suberbiola, A. Perez-Garcia, J.-C. Rage, and P. Vincent. "Mesozoic Marine Palaeobiogeography in Response to Drifting Plates." *Gondwana Research* 26, nos. 3–4 (2014): 869–87.

Bardet, N., E. Gheerbrant, A. Noubhani, H. Cappetta, S. Jouve, E. Bourdon, X. Pereda Suberbiola, et al. "Les Vertébrés des phosphates crétacés-paléogènes du Maroc" [The vertebrates from the Cretaceous-Palaeogene (72.1–47.8 Ma) phosphates of Morocco]. In *Paléontologie des Vertébrés du Maroc, État des connaissances* [Vertebrate paleontology of Morocco: The state of knowledge], edited by S. Zouhri, 351–452. Paris: la Société Géologique de France, 2017.

Bardet, N., A. Houssaye, J. C. Rage, and X. Pereda Suberbiola. "Marine Squamate Radiation during the Cenomanian-Turonian (Late Cretaceous): The Role of the Mediterranean Tethys." *Bulletin de la Société Géologique de France* 179, no. 6 (2008): 605–22.

Bardet, N., A. Houssaye, P. Vincent, X. Pereda Suberbiola, M. Amaghzaz, E. Jourani, and S. Meslouh. "Mosasaurids (Squamata) from the Maastrichtian Phosphates of Morocco: Biodiversity, Palaeobiogeography and Palaeoecology Based on Tooth Morphoguilds." *Gondwana Research Special Issue* 27 (2015): 1068–78.

Benton, M. J., Q. Zhang, S. Hu, Z.-Q. Chen, W. Wen, J. Liu, J. Huang, C. Zhou, T. Xie, J. Tong, and B. Choo. "Exceptional Vertebrate Biotas from the Triassic of China, and the Expansion of Marine Ecosystems after the Permo-Triassic Mass Extinction." *Earth-Science Reviews* 125 (2013): 199–243.

Bernard, A., C. Lecuyer, P. Vincent, R. Amiot, N. Bardet, E. Buffetaut, G. Cuny, F. Fourel, F. Martineau, and J.-M. Mazin. "Regulation of Body Temperature by Some Mesozoic Marine Reptiles." *Science* 328 (2010): 1379–82.

Braun, J., and W. E. Reif. "Konstruktionsmorphologie, Nr. 169: A Survey of Aquatic Locomotion in Fishes and Tetrapods." *Neues Jahrbuch für Geologie und Paläontologie. Abhandlungen* 169, no. 3 (1985): 307–32.

Cadena, E. A., and J. F. Parham. "Oldest Known Marine Turtle? A New Protostegid from the Lower Cretaceous of Colombia." *PaleoBios* 32 (2015): 1–42.

Chen, X. H., R. Motani, L. Cheng, D. Y. Jiang, and O. Rieppel. "A Carapace-Like Bony 'Body Tube' in an Early Triassic Marine Reptile and the Onset of Marine Tetrapod Predation." *PLoS ONE* 9, no. 4 (2014): e94396.

Cheng, L., X. H. Chen, Q. H. Shang, and X. C. Wu. "A new Marine Reptile from the Triassic of China, with a Highly Specialized Feeding Adaptation." *Naturwissenschaften* 101, no. 3 (2014): 251–59.

Chun, L., O. Rieppel, C. Long, and N. Fraser. "The Earliest Herbivorous Marine Reptile and Its Remarkable Jaw Apparatus." *Science Advances* 2, no. 5 (2016): e1501659.

Crofts, S., J. Neenan, T. Scheyer, and A. Summers. "Tooth Occlusal Morphology in the Durophagous Marine Reptiles, Placodontia (Reptilia: Sauropterygia)." *Paleobiology* 43 (2017): 114–28.

Cuvier, G. "Sur le Grand Animal Fossile des Carrières de Maestricht." *Annales du Muséum d'Histoire Naturelle Paris* 12 (1808): 145–76.

Ewin, T., and B. Thuy. "Brittle Stars from the British Oxford Clay: Unexpected Ophiuroid Diversity on Jurassic Sublittoral Mud Bottoms." *Journal of Paleontology* 91 (2017): 781–98.

Hagdorn, H., X. Wang, and C. Wang. "Palaeoecology of the Pseudoplanktonic Triassic Crinoid *Traumatocrinus* from Southwest China." *Palaeogeography, Palaeoclimatology, Palaeoecology* 247 (2007): 181–196.

Houssaye, A., J. Lindgren, R. Pellegrini, A. H. Lee, D. Germain, and M. J. Polcyn. "Microanatomical and Histological Features in the Long Bones of Mosasaurine Mosasaurs (Reptilia, Squamata)—Implications for Aquatic Adaptation and Growth Rates." *PLoS ONE* 8, no. 10 (2013): e76741.

Jouve, S., M. Iarochène, B. Bouya, and M. Amaghzaz. "A New Species of *Dyrosaurus* (Crocodylomorpha, Dyrosauridae) from the Early Eocene of Morocco: Phylogenetic Implications." *Zoological Journal of the Linnean Society* 148, no. 4 (2006): 603–56.

Kardong, K. V. *Vertebrates: Comparative Anatomy, Function, Evolution*. New York: McGraw-Hill, 1995.

Kear, B. P. "Marine Reptiles from the Lower Cretaceous of South Australia: Elements of a High-Latitude Cold-Water Assemblage." *Palaeontology* 49 (2006): 837–56.

Kear, B., N. I. Schroeder, and M. S. Y. Lee. "An Archaic Crested Plesiosaur in Opal from the Lower Cretaceous High-Latitude Deposits of Australia." *Biology Letters* 2 (2006): 615–19.

Konishi, T., M. W. Caldwell, T. Nishimura, K. Sakurai, and K. Tanoue. "A New Halisaurine Mosasaur (Squamata: Halisaurinae) from Japan: The First Record in the Western Pacific Realm and the First Documented Insights into Binocular Vision in Mosasaurs." *Journal of Systematic Palaeontology* 14, no. 10 (2015): 809–839.

Lindgren, J., M. J. Polcyn, and B. A. Young. "Landlubbers to Leviathans: Evolution of Swimming in Mosasaurine Mosasaurs." *Paleobiology* 37, no. 3 (2011): 445–69.

Liu, S., A. S. Smith, Y. Gu, J. Tan, C. K. Liu, and G. Turk. "Computer Simulations Imply Forelimb-Dominated Underwater Flight in Plesiosaurs." *PLoS Computational Biology* 11, no. 12 (2015): e1004605.

Massare, J. A. "Tooth Morphology and Prey Preference of Mesozoic Marine Reptiles." *Journal of Vertebrate Paleontology* 7 (1987): 121–37.

Mcgowan, C., and R. Motani. *Handbook of Paleoherpetology. Part. 8: Ichthyopterygia*. Munich: Dr. Friedrich Pfeil, 2003.

Modesto, S. P., and J. S. Anderson. "The Phylogenetic Definition of Reptilia." *Systematic Biology* 53, no. 5 (2004): 815–21.

Moreira, C. A., D. W. Dempster, and R. Baron. "Anatomy and Ultrastructure of Bone—Histogenesis, Growth and Remodeling." Endotext [Internet], 2019.

Motani, R. "Evolution of Fish-Shaped Reptiles (Reptilia: Ichthyopterygia) in Their Physical Environments and Constraints." *Annual Review of Earth and Planetary Sciences* 33 (2005): 395–420.

Motani, R., D.-Y. Jiang, A. Tintori, O. Rieppel, and G.-B. Chen. "Terrestrial Origin of Viviparity in Mesozoic Marine Reptiles Indicated by Early Triassic Embryonic Fossils." *PLoS ONE* 9, no. 2 (2014): 88640.

Motani, R., B. M. Rothschild, and W. Wahl. "Large Eyeballs in Diving Ichthyosaurs." *Nature* 402 (1999): 747.

Nagashima, H., F. Sugahara, M. Takechi, R. Ericsson, Y. Kawashima-Ohya, Y. Narita, and S. Kuratani. "Evolution of the Folding and Creation of New Muscle Connections." *Science* 325 (2009): 193–96.

Rage, J. C., and F. Escuillié. "Un nouveau serpent bipède du Cénomanien (Crétacé). Implications phylétiques." *Comptes Rendus de l'Académie des Sciences—Series IIA—Earth and Planetary Science* 330, no. 7 (2000): 513–20.

Renesto, S., and R. Stockar. "Exceptional Preservation of Embryos in the Actinopterygian *Saurichthys* from the Middle Triassic of Monte San Giorgio, Switzerland." *Swiss Journal of Geosciences* 102 (2009): 323–30.

Rieppel, O. "The Cranial Anatomy of *Placochelys placodonta Jaekel*, 1902, and a Review of the Cyamodontoidea (Reptilia, Placodonta)." *Fieldiana: Geology, New Series* 45 (2001): 1–104.

Röhl, H.-J., A. Schmid-Röhl, W. Oschmann, A. Frimmel, and L. Schwark. "The Posidonia Shale (Lower Toarcian) of SW-Germany: An Oxygen-Depleted Ecosystem Controlled by Sea Level and Palaeoclimate." *Palaeogeography, Palaeoclimatology, Palaeoecology* 165, nos. 1–2 (2001): 27–52.

Russell, D. "Systematics and Morphology of American Mosasaurs." *Bulletin of the Peabody Museum of Natural History* 23 (1967).

Sato, T., L.-J. Zhao, X.-C. Wu, and C. Li. "A New Specimen of the Triassic Pistosauroid *Yunguisaurus*, with Implications for the Origin of Plesiosauria (Reptilia; Sauropterygia)." *Palaeontology* 57 (2013): 55–76.

Schoch, R. R., and H.-D. Sues. "A Middle Triassic Stem-Turtle and the Evolution of the Turtle Body Plan." *Nature* 523 (2015): 584–87.

Sharp, Z. D., V. Atudorei, and H. Furrer. "The Effect of Diagenesis on Oxygen Isotope Ratios of Biogenic Phosphates." *American Journal of Science* 300 (2000): 222–37.

Wegener, A. *Die Entstehung der Kontinente und Ozeane*. F. Vieweg & Sohn Ed, 1915.

White, M. A., A. G. Cook, S. A. Hocknull, T. Sloan, G. H. Sinapius, and D. A. Elliott. "New Forearm Elements Discovered of Holotype Specimen *Australovenator wintonensis* from Winton, Queensland, Australia." *PLoS ONE* 7, no. 6 (2012): e39364.

RESOURCES

Books

Berta, A., J. L. Sumich, and K. M. Kovacs. *Marine Mammals: Evolutionary Biology*. 3rd ed. Amsterdam: Academic Press, 2005 (2015).

Berta, A. *The Rise of Marine Mammals: 50 Million Years of Evolution*. Baltimore: John Hopkins University Press, 2017.

Buffetaut, E. *Des fossiles et des hommes*. Paris: Robert Laffont, 1991.

Cardot, C. *Georges Cuvier. La révélation des mondes perdus*. Besançon, France: Sekoya, 2009.

Cuny, G., and A. Bénéteau. *Requins. De la préhistoire à nos jours*. Paris: Belin, 2009.

Deconinck, J.-F., *Paléoclimats. L'enregistrement des variations climatiques*. Paris: Vuibert, 2006.

Everhart, M. J. *Oceans of Kansas: A Natural History of the Western Interior Sea*. 2nd ed. Bloomington: Indiana University Press, 2005.

Mcgowan, C. *Dinosaurs, Spitfires and Sea Dragons*. Harvard, MA: Harvard University Press, 1992.

Rieppel, O. *Mesozoic Sea Dragons—Triassic Marine Life from the Ancient Tropical Lagoon of Monte San Giorgio*. Bloomington: Indiana University Press, 2019.

Sharpe, T. *The Fossil Woman: A Life of Mary Anning*. Wimborne Minster, Dorset, UK: The Dovecote Press, 2020.

Steyer, J.-S., and A. Bénéteau. *La Terre avant les dinosaures*. Paris: Belin, 2009.

Taquet, P. *Georges Cuvier. Naissance d'un génie*. Paris: Odile Jacob, 2006.

Vincent, P., and G. Suan. *Les fossiles*. Plouédern, France: Gisserot, 2011.

Newspapers and Periodicals

"Les mers au Mésozoïque." *Espèces*, hors-série, no. 2 (2016).

"Sites paléontologiques de France." *Espèces*, no. 28 (2018).

"Les falaises des Vaches-Noires, un gisement emblématique du Jurassique à Villers-sur-Mer, Normandie." *Fossiles*, hors-série, no. 4 (2013).

Novel

Chevalier, T. *Prodigieuses creatures* (reedition). Paris: Gallimard, coll. "Folio" no 5267, 2011.

Films

Cuissot, P. *Au temps des dinosaures*. Documentary. Bonne Pioche Production, 2021.

Vuong, P., and R. Chapalain. *Océanosaures, voyage au temps des dinosaures*. Documentary. N3DLand Productions, 2010.

Internet Sites

Oceans of Kansas Paleontology, www.oceansofkansas.com

The Plesiosaur directory, www.plesiosauria.com

For Children

Allain, R., A. Alter, and B. Perroud. *Les dinosaures, ce qu'on ne sait pas encore*. Paris: Le Pommier, 2016.

Aumont, A., and Miniac. *Léon aux falaises des Vaches-Noires*. Hérouville-Saint-Clair, France: OREP/Paléospace, 2011.

Smith, A. S., J. Emmett, and A. Larkum. *The Plesiosaur's Neck*. Preston, Lancashire, UK: UCLan Publishing, 2021.

Vincent, P., and Camouche. *Estelle et Noé: A la découverte des dinosaures*. Vincennes, France: Millepages, 2020.

ACKNOWLEDGMENTS

We would like to thank the following individuals and institutions who graciously and generously provided several references and illustrations:

Allemand, Rémi (Department of Anthropology, University of Toronto, Canada)

Buffetaut, Eric (CNRS, École normale supérieure, Paris, France)

Cazes, Lilian (CR2P—Centre de Recherche en Paléontologie—Paris, CNRS, MNHN, Paris, France)

Chapman, Sandra (The Natural History Museum, London, United Kingdom)

Chun, Li (Institute of Vertebrate Paleontology and Paleoanthropology, Beijing, China)

Colin, Cécile (Galerie de Paléontologie et d'Anatomie comparée, MNHN, Paris, France)

Cuny, Gilles (Laboratoire d'Écologie des hydrosystèmes naturels et anthropisés, Université Claude Bernard Lyon 1, Villeurbanne, France)

Cornejo, Manuel (Association "Les Amis de Maurice Ravel," Saint-André-lez-Lille, France)

Dejouannet, Jean-François (UMS2700–2AD, Galerie de Paléontologie et d'Anatomie comparée, MNHN, Paris, France)

El Hammoumi Idrissi, Aïcha (OCP Group, Casablanca, Morocco)

Falconnet, Jocelyn (MNHN, Paris, France)

Fernandez, Sophie (CR2P—Centre de Recherche en Paléontologie—Paris, MNHN, Paris, France)

Hauff, Rolf (Urweltmuseum Hauff, Holzmaden, Germany; www.urweltmuseum.de)

Jacquot-Haméon, Justine (MNHN, Paris, France)

Jagt, John (Natuurhistorisch Museum Maastricht, Maastricht, The Netherlands)

Kear, Benjamin (Museum of Evolution, Uppsala University, Uppsala, Sweden)

Khadiri Yazami, Oussama (OCP Group, Khouribga, Morocco)

Khaldoune, Fatima (OCP Group, Khouribga, Morocco)

Klug Christian (Paläontologisches Institut und Museum, Universität Zürich, Switzerland)

Lemzaouda, Christian (CR2P—Centre de Recherche en Paléontologie—Paris, CNRS, MNHN, Paris, France)

Letenneur, Charlène (CR2P—Centre de Recherche en Paléontologie—Paris, MNHN, Paris, France)

Loubry, Philippe (CR2P—Centre de Recherche en Paléontologie—Paris, CNRS, MNHN, Paris, France)

Marozsán, Imre (Letter Product Management Directorate, Philatelic Department, Budapest, Hungary)

Martill, David (School of the Environment, Geography and Geosciences, University of Portsmouth, Portsmouth, United Kingdom)

Maxwell, Erin (Staatliches Museum für Naturkunde, Stuttgart, Germany)

Muizon, Christian de (CR2P—Centre de Recherche en Paléontologie—Paris, CNRS, MNHN, Paris, France)

Ősi, Attila (MTA-ELTE, Lendület Dinosaur Research Group, Hungarian Academy of Sciences, Budapest, Hungary)

Pérez Garcia, Adán (Universidad Nacional de Educación a Distancia, Madrid, Spain)

Rieber, Hans (Universität Zürich, Switzerland)

Rieppel, Olivier (Section of Earth and Planetary Sciences, The Field Museum, Chicago, USA)

Roux, Aurélie (Photothèque / Vidéothèque, MNHN, Paris, France)

Scheyer, Torsten (Paläontologisches Institut und Museum, Universität Zürich, Switzerland)

Sciau, Laurent (Millau, France)

Sciau, Jacques (Musée municipal de Millau, France)

Tong, Haiyan (Palaeontological Research and Education Centre, Mahasarakham University, Maha Sarakham, Thailand)

ILLUSTRATION CREDITS

• **For all the illustrations contained in this work:** © **Alain Bénéteau/2021,** except for p. 54, bottom (Jean-François Dejouannet); p. 84, top (Charlène Letenneur); and p. 192 (Justine Jacquot-Haméon).
• **Infographics and templates: Thomas Haessig,** except for p. 90, top (Sophie Fernandez).
The complete references for the data provided can be found on pp. 203–5.
Abbreviations: CR2P: Centre de Recherche en Paléontologie, Paris; CNRS: Centre national de la recherche scientifique, Paris; IVPP: Institute of Vertebrate Paleontology and Paleoanthropology, Beijing; MNHN: Muséum national d'Histoire naturelle, Paris; NHMM: Natuurhistorisch Museum Maastricht; SMNS: State Museum of Natural History Stuttgart.

p. 9, top left: iStock/Ken Griffiths; p. 9, top right, iStock/Rawlinson-Photography; p. 9, bottom left: iStock/Kunhui Chih; p. 9, bottom right: iStock/Tammy616; p. 21: Stéphane Jouve/with the kind permission of University of the Sorbonne; p. 27: data quoted from Modesto and Anderson (2004); pp. 30–31: data quoted from Bardet et al. (2014); p. 34: data quoted from Motani (2005); p. 37: data quoted from Motani et al. (1999); p. 38: data quoted from McGowan and Motani (2003); p. 40: © Science Photo Library/Akg-images; p. 41: © Nathalie Bardet; p. 43: data quoted from Bardet (2014); p. 44: data quoted from Croft et al. (2017) and Rieppel (2001); p. 48: Wikimedia/CC BY-SA 3.0/Ninjatacoshell; p. 50, top: © Peggy Vincent/ with the kind permission of University of the Sorbonne; p. 50, left bottom: data quoted from Sato et al. (2013); p. 54, top: © Peggy Vincent/ with the kind permission of MNHN; p. 54, bottom: © Jean-François Dejouannet/UMS 2700 CNRS/MNHN; p. 61, right: data quoted from Chen et al. (2014); p. 63: data quoted from Bardet et al. (2014); p. 64, bottom: © Peggy Vincent/ with the kind permission of MNHN; p. 67, right top: data quoted from Bardet et al. (2014); p. 68, bottom: © Peggy Vincent/ with the kind permission of MNHN; p. 69: © Stéphane Jouve/ with thc kind permission of MNHN; p. 70: Wikimedia/CC BY-SA 4.0/Ghedoghedo; p. 74: © Stéphane Jouve; p. 75, right: data quoted from Bardet et al. (2014); p. 77 and p. 78, top: data quoted from Nagashima et al. (2009); p. 78, bottom: data quoted from Schoch and Sues (2015); p. 80: data quoted from Cadena and Parham (2015); p. 81: Wikimedia/CC BY-SA 4.0/Ghedoghedo; p. 84, top: © Christian Lemzaouda/CR2P/CNRS/MNHN/ reconstruction by Charlène Letenneur/CR2P/MNHN; p. 84, bottom: © Nathalie Bardet/ with the kind permission of NHMM; p. 86, top, and p. 87: data quoted from Bardet et al. (2014) and from Bardet et al. (2008); p. 86, bottom: data quoted from Russell (1967); p. 89, left: © Alexandra Houssaye; p. 89, right: data quoted from Rage and Escuillié (2000); p. 90, top: © Sophie Fernandez, CR2P/MNHN; p. 91: data quoted from Houssaye et al. (2013); p. 91, bottom: data quoted from Lindgren et al. (2011); p. 92: © Nathalie Bardet; p. 94: © Philippe Loubry/CR2P/CNRS/MNHN and data quoted from Bardet et al. (2015); p. 95: Wikimedia/Jan Arkesteijn; p. 96: © Nathalie Bardet and data quoted from Cuvier (1808); p. 97, top: © Nathalie Bardet; p. 97, bottom: © Nathalie Bardet with the kind permission of NHMM; p. 101: © Philippe Loubry/CR2P/CNRS/MNHN; p. 102: © Peggy Vincent/ with the kind permission of MNHN; p. 105, bottom: data quoted from Wegener (1915); p. 110: © Hungarian Post (Magyar Posta); p. 113: © Hans Rieber (Universität Zürich, Switzerland); p. 114: data quoted from Renesto and Stockar (2009) and Sharp et al. (2000); p. 118: data quoted from Braun and Reif (1985); p. 122, top: © Li Chun (IVPP, Beijing, China); p. 122, bottom: data quoted from Benton et al. (2013); p. 123: data quoted from Hagdorn et al. (2007); p. 129: data quoted from Kardong (1995) (middle) and Moreira et al. (2019) (right); pp. 130–131: © Alexandra Houssaye; p. 132: data quoted from Cheng et al. (2014); p. 133: data quoted from Chun et al. (2016); p. 139, top: © Urweltmuseum/www.urweltmuseum.de; p. 139, bottom: data quoted from Röhl et al. (2001); p. 141: © Peggy Vincent/with the kind permission of MNHN; p. 142: © With the kind permission of SMNS (photography U. Schmid); p. 143: data quoted from Motani et al. (2014); p. 148: data quoted from Motani (2005); p. 149: data quoted from Liu et al. (2015); p. 150, bottom: data quoted from Konishi et al. (2015); p. 151, top: © Rémi Allemand (University of Toronto, Canada) and data quoted from Allemand et al. (2019); p. 152: © David Martill (University of Portsmouth, United Kingdom); p. 153: data quoted from Ewin and Thuy (2017); p. 154: © Alexandra Houssaye; p. 155: data quoted from Bernard et al. (2010); p. 164, top: © Nathalie Bardet; p. 164, bottom: © Benjamin Kear (University of Uppsala, Sweden); p. 165: data quoted from White et al. (2012); p. 169: data quoted from Kear et al. (2006); p. 171: © Benjamin Kear (University of Uppsala, Sweden) and data quoted from Kear (2006); p. 172: © Nathalie Bardet; p. 173: data quoted from Bardet et al. (2017); p. 174, left: © Philippe Loubry /CR2P/ CNRS/MNHN; p. 175: © Nathalie Bardet; p. 181: data quoted from Massare (1987); p. 184: © Nathalie Bardet; p. 192: © Justine Jacquot-Haméon/MNHN.

First published in France as *La mer au temps des dinosaures*.

Published by Princeton University Press
41 William Street, Princeton, New Jersey 08540
99 Banbury Road, Oxford OX2 6JX
press.princeton.edu

ISBN 9780691243948
ISBN (e-book) 9780691243993

British Library Cataloging-in-Publication Data is available

Editorial: Robert Kirk and Megan Mendonça
Production Editorial: Natalie Baan
Text Design for English edition: D & N Publishing, Wiltshire, UK
Cover Design: Wanda España
Production: Ruthie Rosenstock
Publicity: Caitlyn Robson and Matthew Taylor
Copyeditor: Will DeRooy

Jacket illustration by Alain Bénéteau

This book has been composed in CMU Bright (main text) and Good Pro Condensed (captions)

Printed on acid-free paper. ∞

Printed in China

10 9 8 7 6 5 4 3 2 1